T0280606

Robert E. ROBERSON
Departement of Aerospace and Mechanical Engineering Sciences
University of California, San Diego

Pierre Y. WILLEMS
Institut de Mécanique
Université de Louvain

Jens WITTENBURG
Lehrstuhl und Institut für Mechanik
Technische Hochschule Hannover

INTERNATIONAL CENTRE FOR MECHANICAL SCIENCES

COURSES AND LECTURES - No. 102

ROBERT ROBERSON
UNIVERSITY OF CALIFORNIA, SAN DIEGO
PIERRE WILLEMS
UNIVERSITÉ CATHOLIQUE, LOUVAIN
JENS WITTENBURG
TECHNISCHE UNIVERSITÄT, HANNOVER

ROTATIONAL DYNAMICS OF ORBITING GYROSTATS

DEPARTMENT OF GENERAL MECHANICS
COURSE HELD IN DUBROVNIK
SEPTEMBER 1971

UDINE 1971

SPRINGER-VERLAG WIEN GMBH

Originally published by Springer-Verlag Wien New York in 1972

ISBN 978-3-211-81198-6 ISBN 978-3-7091-2930-2 (eBook)

DOI 10.1007/978-3-7091-2930-2

PREFACE

At the suggestion of Prof. Sobrero of CISM I organized a series of lectures to be presented by me and several colleagues at Dubrovnik in September 1971 under the joint auspices of CISM and the University of Zagreb. For his encouragement and support, I wish to immediately express my thanks.

The lectures were organized in two series, and three hours of lectures were presented in each series each day during 13 - 17 September. This book contains the lectures of the first series, given by the undersigned, Dr. Pierre Y. Willems and Dr. Jens Wittenburg. Each series was devoted to one aspect of special current importance relating to the rotational behaviour of spacecraft.

The subject of this first series was the rotational dynamics of orbiting gyrostats. Two topics were emphasized, for these two underlie the major technological applications involving spinning gyrostatic spacecraft and gyrostatic spacecraft in equilibrium states under the action of gravitational torque. The first problem is to establish the conditions under which a gyrostat can spin freely in space without being subject to any torque, and to develop complete solutions in terms of elliptic functions for the spinning motion of a special kind of gyrostat in which the in-

ternal angular momentum is aligned with a principal axis of inertia. The second problem is to establish the nature of a gyrostat's gravitational equilibrium orientations and their stability. Technological opplications build on these problems in pure rotational dynamics by introducing dissipative mechanisms, a subject not covered in these lectures. However, as intermediate ground between this series of lectures and the companion one on spacecraft flexibility, the effect of flexibility on the stability of gravitational equilibria is included here.

Dr. Wittenburg prepared and presented Lectures 6 and 7, Dr. Willems Lectures 11 and 12, and I the remaining lectures

RE Roberson

Dubrovnik, September 1971

1. Introductory Remarks

The subject of rotational dynamics was born in mid-18th Century. d'Alembert and Euler both solved the problem of the precession of the equinoxes in 1749, a case in which it is enough to know how to handle an infinitesimal rotation. At that time, nothing was known about how to handle a general rotation.

In 1750, Euler published his general theory that can be said to mark the birth of the subject, following it in 1758 and 1760 by solutions of the problem of a free body spinning in an arbitrary way about a fixed point. It is here that he introduced the "Euler angle" representation of rotation into mechanics.

During the three decades, 1750-1780, d'Alembert published a number of inconsequential works on the subject, Lagrange took up the problem of the libration of the moon (1764 and 1780) , and Euler himself codified the subject in his two volume treaties on rotation in 1765 and added a four-parameter representation of rotation in 1770. Finally, in his Mécanique Céleste of 1788, Lagrange gave the solution to his top problem.

Of all these works, those on the free spinning

body and on libration in the gravitational field are of greatest importance to us here, because they are the forerunners and analogues of the problems discussed here for gyrostats.

By 1788 our subject had been born, matured, and died. Or, at least, had almost died. It held to life with only the most tenuous thread for the next hundred years. Some interest but little real progress was generated by the discovery of Kovalevskaya's general case of top motion and of a number of singular solutions to that problem. Astronomy was, in a sense, an application area, but nothing significantly new was generated there. The 19th Century saw mainly the work of the 18th Century redone in "modern" terms.

We must, however, single out two important memoirs that marked the birth of the theory of rotating gyrostats. One of these was by Wangerin in 1889: Motivated purely as a generalization of the free rigid body, it was published in an obscure and relatively unobtainable source, and fell into oblivion until recently. The other, by V. Volterra in 1895 (actually a codification and summary of several of his other papers), was motivated by his attempt to explain the observed variation in latitude on points of the earth by modeling the earth as a gyrostat. In neither case is the motivation aught but intellectual curiosity or "pure" science.

Technological demands in rotational dynamics during the 19th Century were negligible. Serson had constructed a gyroscopic horizon about 1742, and von Bohnenberger had invented the (two-axis) gyroscope in about 1817, but the gyroscope was not popularized until 1852 when Foucault used a single-axis instrument to demonstrate the rotation of the earth. The modest need for gyroscope theory during the 1850-1880 period were adequately served by symmetrical top theory. No real technological applications were made in the 19th Century except to torpedo guidance, to the French Navy's modification of Serson's horizon, and in a very preliminary way to the marine gyrocompass.

During the early 20th Century, technological applications to gyroscopic sensing devices grew, and some theoretical growth accompanied them. Although the classic treatise of Routh (1860) had addressed only the theory - his illustrative examples were "made up" problems - by the time Klein and Sommerfeld's great work appeared (1897-1910) we find substantial attention to applications. Indeed, their entire fourth volume is given to them. Good work was done by several authors early in the century, but it was not copious and no one author stands out as a major contributor to the subject.

By mid-20th Century, gyroscopic sensing devices had become more accurate, more complicated, and were applied under more diverse physical conditions. Theory responded mainly as the theory of linear automatic control systems incorporating linear (small output angle) gyros. Since 1952, however, a new application area - spacecraft - has rejuvenated the subject of rotational dynamics more than any other single factor since its birth. Now a technological situation exists in which large-angle, multi-body problems are the norm, not artificial rarities. The rejuvenation has been possible, of course, only because of the rise of the computer at just the time it was needed.

In the late 1960's, spinning gyrostats began to assume special practical importance because of their close relationship to the so-called dual-spin systems used for passive spin-stabilization of certain communication satellites. Since then, a number of the classical problems for the rigid body have been reworked for the gyrostat. These lectures are a report of some of the results.

The series has a natural grouping into three sets of lectures. The first is concerned with the torque-free spinning gyrostat. That system is important both inherently, because some real spacecraft can be modeled very closely by it, and as a starting point for the analysis of more general cases by perturbation methods. The second group concerns the gyrostat in equilibrium orientations in the earth's gravitational field. It is a generalization of Lagrange's libration problem that finds application to passive, gravitationally stabilized spacecraft. The third group extends the latter problem by taking account of body flexibility, thereby providing a tie with the parallel lecture series on the "Dynamics of Flexible Spacecraft."

2. Notation and Mathematical Conventions

Vectors and Dyadics

The mathematics used in rotational dynamics is sim ple enough, and undoubtedly very familiar to the audience. However, the best background does not necessarily imply familiarity with all the notational conventions that will be used here, so it is well ro review them briefly.

Physical space is modeled as a mathematical three-dimensional vector space on the field of real numbers, a normed euclidean space. In many contexts a variety of vectors and vector bases for the space are of simultaneous concern. In resolving a vector into its components with respect to a vector basis in the space, the resolution depends not on the origin of the basis but only on its orientation. Any basis can be translated freely to another origin in our eucliedean space without changing the definition of vector components.

Vectors as inarients are to be distinguished carefully from their representations. The conventional symbol of a vector is a single symbol, underlined: thus $\underline{v}$ represents a vector. Bases used in this work are exclusively orthonormal, consisting of triads of vectors each of unit length, mutually orthogonal, arranged in a dextral (right-hand) sense. The set $\{\underline{e}_\alpha\}$ $(\alpha = 1,2,3)$ represents such a basis. An arbitrary vector

$\underline{v}$ in the space can be represented as the linear form

$$\underline{v} = v_1 \underline{e}_1 + v_2 \underline{e}_2 + v_3 \underline{e}_3 \tag{1}$$

The v_1, v_2, v_3 (briefly, v_α) are scalar constants from the field of real numbers, called the components of the vector with respect to the basis $\{\underline{e}_\alpha\}$. They are unique for any vector, given the basis. We often will write the components in a column matrix, using the notation $v = [v_\alpha]$ for it.

The summation convention of cartesian tensor analysis is used. Repeated indices, those that appear exactly twice in a product, are assumed to be summed over their range without explicitly writing a summation sign. Thus Eq. 1 can be written as $\underline{v} = \sum_{\alpha=1} v_\alpha \underline{e}_\alpha$ or simply as $\underline{v} = v_\alpha \underline{e}_\alpha$. Greek lower case indices have the range 1,2,3 throughout. A summed index also is called a dummy index, and it can be replaced freely by any other suitable letter. A single index letter never can appear more than twice in an expression without using an explicit summation sign. If a typical term rather than a sum is intended, the phrase "not summed on ..." is added . An index that appears only once in each additive term of an equation is called a free index. It, too, can be changed to any other letter if the change is made in each term of the equation. In regular tensor analysis each dummy index must appear once as a superscript and once as a subscript (contravariant and covariant components), but with cartesian tensors no distinction is made and both indices are written in the subscript position.

The reader is assumed familiar with the notation of Gibbsian vector analysis. Thus for the vector of a basis we have the scalar ("dot") and vector ("cross") products respectively:

$$\underline{e}_\alpha \cdot \underline{e}_\beta = \delta_{\alpha\beta} \qquad \underline{e}_\alpha \cdot \underline{e}_\beta = \varepsilon_{\alpha\beta\gamma}\,\underline{e}_\gamma \tag{2ab}$$

Here $\delta_{\alpha\beta}$, called the Kronecker delta, is defined as $\delta_{\alpha\beta} = 1$ $(\alpha = \beta)$ and $\delta_{\alpha\beta} = 0$ $(\alpha \neq \beta)$. The three-index symbols $\varepsilon_{\alpha\beta\gamma}$ are called epsilon symbols of tensor analysis and are defined as

$$\varepsilon_{\alpha\beta\gamma} = \begin{cases} +1 & \text{if } (\alpha\beta\gamma) \text{ a cyclic permutation of } (123) \\ -1 & \text{if } (\alpha\beta\gamma) \text{ a cyclic permutation of } (132) \\ 0 & \text{if any two indices equal} \end{cases} \tag{3}$$ (*)

(Cyclic permutations of any triple of numbers (abc) are (abc),(bca) and (cab).) The "right-handedness" of the frame is implicitly defined by the form of Eq. 2b.

It follows that if $\underline{u} = u_\alpha \underline{e}_\alpha$ and $\underline{v} = v_\alpha \underline{e}_\alpha$,

---

(*) Some useful properties are:

$$\varepsilon_{\alpha\beta\gamma} = \varepsilon_{\gamma\alpha\beta} = \varepsilon_{\beta\gamma\alpha} = -\varepsilon_{\alpha\gamma\beta} = -\varepsilon_{\beta\alpha\gamma} = -\varepsilon_{\gamma\beta\alpha}$$

$$\varepsilon_{\alpha\beta\gamma}\,\varepsilon_{\lambda\mu\gamma} = \delta_{\alpha\lambda}\delta_{\beta\mu} - \delta_{\alpha\mu}\delta_{\beta\lambda}$$

$$\varepsilon_{\alpha\beta\lambda}\,\varepsilon_{\alpha\beta\mu} = 2\delta_{\lambda\mu}$$

(4) $$\underline{u} \cdot \underline{v} = u_\alpha v_\alpha$$

(5a) $$\underline{u} \cdot \underline{v} = u_\alpha v_\beta (e_\alpha \cdot \underline{e}_\beta) = u_\alpha v_\beta \varepsilon_{\alpha\beta\gamma} \underline{e}_\gamma = (\varepsilon_{\gamma\alpha\beta} u_\alpha) v_\beta \underline{e}_\gamma$$

It is very convenient to define, for an arbitrary set of vector components u_α, a two-index symbol $\tilde{u}_{\lambda\mu}$ by

(6) $$\tilde{u}_{\lambda\mu} = \varepsilon_{\lambda\alpha\mu} u_\alpha$$

Thus Eq. 5a can be rewritten

(5b) $$\underline{u} \cdot \underline{v} = (\tilde{u}_{\lambda\beta} v_\beta) \underline{e}_\lambda$$

Let $\{\underline{e}^1_\alpha\}$ and $\{\underline{e}^2_\alpha\}$ be two different sets of orthonormal base vectors. The superscripts are simply base labels, while the subscripts label the individual vectors that comprise a basis. Later the base label will be identified with a material body label by visualizing the several bases as rigidly embedded in the various bodies that comprise the material system of interest, or as specific orientation reference frames. By Eq. 1, any one of the $\underline{e}^2_\alpha$ can be written as a linear sum of the $\underline{e}^2_\alpha$ e.g. $\underline{e}^2_1 = A_{11} \underline{e}^1_1 + A_{12} \underline{e}^1_2 + A_{13} \underline{e}^1_3$. Using the notational conventions of cartesian tensors, the three equations of this type can be written in the compact form $\underline{e}^2_\alpha = A_{\alpha\beta} \underline{e}^1_\beta$. When a number of bases occur in the same context it is convenient to label $A_{\alpha\beta}$ by the indices of the base concerned, put in a superscript position. Thus

$$\underline{e}^{2}_{\alpha} = A^{21}_{\alpha\beta}\, \underline{e}^{1}_{\beta} \tag{7a}$$

By this same convention

$$\underline{e}^{1}_{\lambda} = A^{12}_{\lambda\mu}\, \underline{e}^{2}_{\mu} \tag{7b}$$

To relate the coefficients in Eqs. 7a and 7b, dot-multiply Eq. 7b by $\underline{e}^{2}_{\alpha}$, but where this vector appears on the left, replace it by Eq. 7a. Thus $\underline{e}^{1}_{\lambda}\cdot(A^{21}_{\alpha\beta}\, \underline{e}^{1}_{\beta}) = A^{12}_{\lambda\mu}\, \underline{e}^{2}_{\mu}\cdot \underline{e}^{2}_{\alpha}$. It is easily seen form this that

$$A^{21}_{\alpha\lambda} = A^{12}_{\lambda\alpha} \tag{8}$$

The scalar product $\underline{e}^{2}_{\lambda}\cdot \underline{e}^{1}_{\mu} = A^{21}_{\lambda\mu}$ can be identified as the product of the lengths of these vectors, both unity, by the cosine of the angle between them. For this reason the $A^{21}_{\alpha\beta}$ are called direction cosines.

So far there are no restrictions on the coefficients $A^{21}_{\alpha\beta}$ in Eq. 7a. However, problems in rotational dynamics mainly involve so-called linear orthogonal transformations. In these the orthonormal set $\{\underline{e}^{1}_{\alpha}\}$ is rotated as a rigid unit into the new orthononormal $\{\underline{e}^{2}_{\alpha}\}$, and we must use the orthonormality conditions to establish some restrictions on the admissible constants $A^{21}_{\alpha\beta}$. Formally,

$$\delta_{\alpha\beta} = \underline{e}^2_\alpha \cdot \underline{e}^2_\beta = \left(A^{21}_{\alpha\lambda}\, e^1_\lambda\right)\cdot\left(A^{21}_{\beta\mu}\, \underline{e}^1_\mu\right) =$$

$$= A^{21}_{\alpha\lambda} A^{21}_{\beta\mu}\, \delta_{\lambda\mu}$$

or

$$A^{21}_{\alpha\lambda}\, A^{21}_{\beta\lambda} = \delta_{\alpha\beta} \tag{9}$$

This is a necessary, but not sufficient condition for rigid rotation: it assures that the axes remain orthogonal and of unit length, but does not guarantee that a right-handed frame remain right-handed. To assure this property, one adjoins a condition that the determinant of the 3x3 array constructed from the $A^{21}_{\alpha\beta}$ equal $+1$.

Next, consider the transformation of vector components under a rotation of the basis. Let $\underline{v}$ be resolved into both $\{\underline{e}^1_\alpha\}$ and $\{\underline{e}^2_\alpha\}$ bases according to

$$\underline{v} = v^1_\alpha\, \underline{e}^1_\alpha = v^2_\alpha\, \underline{e}^2_\alpha \tag{10}$$

The superscripts on the components again are labels of the bases with respect to which the components are formed. When there is no danger of confusion, the component superscript can be omitted. Form the scalar product of Eq. 10 by $\underline{e}^1_\lambda$: that is, $v^1_\alpha\, \underline{e}^1_\alpha \cdot \underline{e}^1_\lambda =$ $= v^2_\alpha\, \underline{e}^2_\alpha \cdot \underline{e}^1_\lambda$ or $v^1_\alpha \delta_{\alpha\lambda} = v^2_\alpha A^{21}_{\alpha\lambda}$. Next form the scalar product by $\underline{e}^2_\lambda$. The two results can be collected as

$$\overset{1}{v}_{\lambda} = \overset{2}{v}_{\alpha} \overset{21}{A}_{\alpha\lambda} \qquad \overset{2}{v}_{\lambda} = \overset{21}{A}_{\lambda\alpha} \overset{1}{v}_{\alpha} \tag{11}$$

From Eq. 11b it is seen that the components transform exactly as do the base vectors themselves.

One constructs an invariant called a vector, represented by $\underline{v}$, as a linear form in the base vectors. The coefficients of the linear form are the vector components. In a similar way one can construct an entity that is a quadratic form in the base vectors:

$$\mathsf{T} = t_{\alpha\beta} \underline{e}_{\alpha} \underline{e}_{\beta} \tag{12}$$

The entity is called a tensor of second order, the coefficients $t_{\alpha\beta}$ are the components of the tensor with respect to the basis (often called simply "the tensor") and the total form in which the base vectors are explicit is the dyadic form of the tensor, or simply the dyadic. Dyadics are represented in this work by bold-face sanserif characters. No geometric significance is to be attached to two base vectors standing together as $\underline{e}_{\alpha}\underline{e}_{\beta}$, in distinction to the other kinds of products of vectors, $\underline{e}_{\alpha} \cdot \underline{e}_{\beta}$ and $\underline{e}_{\alpha} \times \underline{e}_{\beta}$. The concept of a vector basis always underlines the definitions of vector and tensor components, but by confining one's attention to these components alone and their manipulation it is possible to avoid ever having to make the basis ex plicit by the notation. Nevertheless, in many physical problems, and in rotational dynamics particularly, is usually is of great practical value to have the base vectors appear explicitly as in Esq. 1 and 12. This is especially true when several distinct bases may be in simultaneous use, as they are in such problems as gyrodynamics, automatic vehicular navigation, and spacecraft attitude control.

The scalar product of a dyadic by a vector is equivalent to a tensor contraction:

$$\underline{v}\cdot T = (v_\lambda \underline{e}_\lambda)\cdot(t_{\alpha\beta}\underline{e}_\alpha\underline{e}_\beta) = v_\lambda(\underline{e}_\lambda\cdot\underline{e}_\alpha)t_{\alpha\beta}\underline{e}_\beta =$$

$$= v_\lambda \delta_{\lambda\alpha} t_{\alpha\beta}\underline{e}_\beta = v_\alpha t_{\alpha\beta}\underline{e}_\beta \tag{13}$$

That is, the scalar product has been formed with the dyadic's base vector standing closest to the dotting vector. The result of the operation, of course, is a vector. The alternative scalar product $T\cdot\underline{v}$ is a vector whose components are the contraction $t_{\alpha\beta}v_\beta$. In general the two products are not equal. The transformation of tensor components under a rotation can be derived analogously to Eq. 10.

If $T = \overset{1}{t}_{\alpha\beta}\overset{1}{\underline{e}}_\alpha\overset{1}{\underline{e}}_\beta = \overset{2}{t}_{\lambda\mu}\overset{2}{\underline{e}}_\lambda\overset{2}{\underline{e}}_\mu$, then

$$\overset{2}{t}_{\lambda\mu} = \overset{21}{A}_{\lambda\alpha}\overset{21}{A}_{\mu\beta}\overset{1}{t}_{\alpha\beta} \tag{14}$$

Matrices

A matrix of arbitrary order is represented here

either by a single symbol or by a symbol for a typical element enclosed within matrix brackets: thus matrix $A = [A_{ij}]$. The first index designates row number, the second column number. The order of the matrix is specified when it is defined, but the matrix notation does not thereafter give aby explicit notice of the order except in the case of 3×3 and 3×1 matrices. (*) Greek lower case subscripts are used as indices for these. Thus $[A_{\alpha\beta}]$ is to be understood to be a 3×3 matrix. The matrix $[A_\alpha]$ is to be understood to be a 3×1 matrix, inasmuch as 1×3 row matrices normally are encountered here only as transposes or column matrices. Under this convention, a combination such as $A_i B_{ij}$ must be interpreted as the j th element of the column matrix $A^T B$ because only a row matrix A^T (not the column matrix A) can precede B in the matrix multiplication symbolyzed by $A_i B_{ij}$.

The unit matrix of any order is denoted E, or if it is essential to indicate that it is of n th order, by E_n. The unit matrix of any order is $E = [\delta_{ij}]$, where the data symbol again represents the Kronecker delta but not limited to a range i,2,3 for its indices.

With any 3×1 column matrix $A = [A_\alpha]$. we can associate a 3×3 skew-symmetric matrix $\tilde{A} = [\tilde{A}_{\alpha\beta}]$ defined as fol-

(*) A Greek index also used in the few cases where a rectangular matrix has three rows or columns. For example, $[A_{\alpha i}]$ is of order 3x (range of i).

lows. Its elements are

(15a) $$\tilde{A}_{\alpha\beta} = \varepsilon_{\alpha\lambda\beta} A_{\lambda}$$

or, written out,

(15b) $$\tilde{A} = \begin{bmatrix} 0 & -A_3 & A_2 \\ A_2 & 0 & -A_1 \\ -A_2 & A_1 & 0 \end{bmatrix}$$

This is exactly the same definition as that given by Eq. 6, implying that if $\tilde{u}_{\lambda\beta}$ in Eq. 5b are interpreted as the elements of a matrix (they need not be, of course) then the components of the cross product $\underline{u} \cdot \underline{v}$ can be written in a 3×1 matrix as the matrix product $\tilde{u}v$.

A very useful concept is that of partitioned matrices. Given a matrix $A = [A_{ij}]$ of order $n \times m$, it can be partitioned into rectangular cells by arbitrary vertical and horizontal lines: e.g.

(16a) $$a = \left[\begin{array}{cc|c|c|c} a_{11} & a_{12} & a_{13} \; \ldots & \ldots & \ldots \; a_{1m} \\ a_{21} & a_{22} & a_{23} \; \ldots & \ldots & \ldots \; a_{2m} \\ \ldots & & \ldots & \ldots & \ldots \\ \hline \ldots & & \ldots & \ldots & \ldots \\ \hline \ldots & & \ldots & \ldots & \ldots \\ a_{n1} & a_{n2} & a_{n3} & & a_{nm} \end{array}\right]$$

Each cell can be labeled as a matrix in its own right, such as

$$A_{11} = \begin{bmatrix} a_{11} & a_{12} \\ a_{21} & a_{22} \\ & \cdots \end{bmatrix}$$ and matrix a then can be

considered a matrix whose elements are themselves matrices:

$$a = \begin{bmatrix} A_{11} & A_{12} & \cdots & A_{1q} \\ A_{21} & A_{22} & \cdots & A_{2q} \\ \cdot & \cdot & \cdot & \cdot \\ A_{p1} & A_{p2} & & A_{pq} \end{bmatrix} \qquad (16b)$$

where $p \leq n$ and $q \leq m$. If two partitioned matrices are multiplied, the matrix product can be formed by the same rule as if the submatrices were scalar elements, provided the individual products of submatrices make sense; i.e. the submatrices have the proper orders, the number of columns of the left ones being equal to the number of rows of the right ones they respectively multiply. Note that in constructing the transpose of a partitioned matrix, the individual submatrices also must be transposed. There is no consistent attempt here to use a uniform notation for partitioned matrices, but often the submatrices are represented by superscript indices while the individual elements of the submatrix are designated by subscripts. Thus we write $a = [a^{ij}]$ with $a^{ij} = [a^{ij}_{pq}]$.

A very important partitioning used in the sequel

is that of the unit matrix. We denote by $E^{(\alpha)}$ the three columns of the 3×3 unit matrix, or more generally by $E^{(i)}$ the columns of unit matrices of other sizes.

The representation of a vector by a matrix is possible in various ways. A convenient one is based on the idea of a vector array, such as

$$\underline{e} = \begin{bmatrix} \underline{e}_1 \\ \underline{e}_2 \\ \underline{e}_3 \end{bmatrix}$$

This is not a matrix in the usual sense, for the elements of a matrix are limited to scalars form some number field, or to matrices in the case of partitioned matrices. However, rules for operations with such vector arrays can be set forth which give them matrix-like characteristics and make it possible to use them very conveniently in the matrix representation of vectors. As regards notation, one has a similar question as for scalars and matrices: does the symbol x denote a real scalar, a complex number, a square matrix, a column matrix, or something else? Except in very limited contexts it usually is not worth trying to introduce a single notation which makes this clear at first glance, a so-called "self-defining" notation. In practice, the burden is on the user' s memory. The same is true for vector arrays: the bold-face symbol $\underline{u}$ could represent a single vector or a vector array, and one knows which only because an explicit definition is given somewhere in the context.

The rules for vector arrays are quickly summarized. Let $\underline{e}$ and $\underline{f}$ be two 3×1 vector arrays, $\underline{e} = [\underline{e}_\alpha]$ and $\underline{f} = [\underline{f}_\alpha]$. The operation of transposition for arrays is as for matrices: $\underline{e}$ is a column array, $\underline{e}^T$ is a row array. We define a sum as $\underline{e} + \underline{f} = [\underline{e}_\alpha + \underline{f}_\alpha]$. If k is a scalar form some field, we define $k\underline{e} = [k\underline{e}_\alpha]$. Furthermore, if $a = [a_{\alpha\beta}]$ is a 3×3 square matrix, we define $a\underline{e} = [a_{\alpha\beta}\underline{e}_\beta]$ and $\underline{e}^T a = [\underline{e}_\beta a_{\beta\alpha}]$. In short, as regards transposition, addition, multiplication by a scalar or a matrix, the vector array satisfies exactly the same rules as if it were a matrix. As one crollary, the transpose rule $(a\underline{e})^T = \underline{e}^T a^T$ holds. These rules can be enlarged to include operations on vector arrays comparable to the "dot" and "cross" product of Gibbsian vector analysis, but these are not essential here.

Using these conventions we can quickly write implicit definitions of matrix components by such expressions as $\underline{u} = \underline{e}^T u$ or $\underline{u} = u^T \underline{e}$ and of tensor components by $T = \underline{e}^T t \underline{e}$, where $t = [t_{\alpha\beta}]$.

Among the most important expressions developed previously are Eqs. 7, 8, 9, 11b and 14. Using matrix and vector array notations these can be rewritten compactly as

$$\underline{e}^2 = A^{21} \underline{e}^1 \tag{18a}$$

$$\underline{e}^1 = A^{12} \underline{e}^1 \tag{18b}$$

$$A^{21} = (A^{12})^T \tag{18c}$$

(18d) $$A^{21} A^{12} = E$$

(18e) $$\upsilon^2 = A^{21} \upsilon^1$$

(18f) $$t^2 = A^{21} t^1 A^{12}$$

Direction Cosine Matrices

Several important properties of the direction cosine matrix should be mentioned. First, we attach physical significance to the row and columns of such a matrix. Consider the orthogonal transformation

(19ab) $$\underline{e}' = A\,\underline{e} \qquad \text{or} \qquad \underline{e} = A^T\,\underline{e}'$$

Partition A into its columns, denoting them respectively by $A^{(\alpha)}$:

(20) $$\begin{bmatrix} A_{11} & A_{12} & A_{13} \\ A_{21} & A_{22} & A_{23} \\ A_{31} & A_{32} & A_{33} \end{bmatrix} = \begin{bmatrix} A_1^{(1)} & A_1^{(2)} & A_1^{(3)} \\ A_2^{(1)} & A_2^{(2)} & A_2^{(3)} \\ A_3^{(1)} & A_3^{(2)} & A_3^{(3)} \end{bmatrix} = \begin{bmatrix} A^{(1)} & A^{(2)} & A^{(3)} \end{bmatrix}$$

Equation 19b can be rewritten

(*) Note that this implies $A = A^{-1}$ for a direction cosine matrix. If a matrix inverse exists it is unique and commutative with the matrix. Thus we have an alternative form of Eq. 9, namely $A^{21}_{\lambda\alpha} \; A^{21}_{\lambda\beta} = \delta_{\alpha\beta}$.

$$\begin{bmatrix} \underline{e}_1 \\ \underline{e}_2 \\ \underline{e}_3 \end{bmatrix} = \begin{bmatrix} A^{(1)T} \\ A^{(2)T} \\ A^{(3)T} \end{bmatrix} \underline{e}' = \begin{bmatrix} A^{(1)T} \underline{e}' \\ A^{(2)T} \underline{e}' \\ A^{(3)T} \underline{e}' \end{bmatrix} \qquad (21a)$$

or

$$\underline{e}_\alpha = A^{(\alpha)T} \underline{e}' \qquad (21b)$$

In other words, $A^{(\alpha)}$ is the 3×1 matrix of components of the "old" vector $\underline{e}_\alpha$ in the "new" frame $\underline{e}'$. We can manipulate with this component matrix just as we can with any matrix of vector components. Because of the unit length of $\underline{e}_\alpha$, we know

$$A^{(\alpha)T} A^{(\alpha)} = A^{(\alpha)}_\lambda A^{(\alpha)}_\lambda = 1 \qquad (22a)$$

That is, the sum of the squares of elements in any column of a direction cosine matrix is 1. Furthermore, the orthogonality of $\underline{e}_\alpha$ and $\underline{e}_\beta \, (\alpha \neq \beta)$ gives

$$A^{(\alpha)T} A^{(\beta)} = A^{(\alpha)}_\lambda A^{(\beta)}_\lambda = 0 \qquad (22b)$$

These equations, of course, are simply the orthonormality conditions (Eqs. 9) in a different guise.

Now consider vector products instead. For example, $\underline{e}_1 \times \underline{e}_2 = \underline{e}_3$. Equation 5b gives an expression for the component matrices of the vectors multiplied. Applying it to this case

(23a)
$$\tilde{A}^{(1)} A^{(2)} = A^{(3)}$$

In a similar way,

(23b)
$$\tilde{A}^{(2)} A^{(3)} = A^{(1)}$$

(23c)
$$\tilde{A}^{(3)} A^{(1)} = A^{(2)}$$

It is useful to write out one, say Eq. 23a, in component form:

$$(\tilde{A}^{(1)})_{\alpha\beta} (A^{(2)})_{\beta} = (A^{(3)})_{\alpha}$$

But $\tilde{A}^{(1)}_{\alpha\beta} = \varepsilon_{\alpha\lambda\beta} A^{(1)}_{\lambda}$, so

(24)
$$A^{(3)}_{\alpha} = \varepsilon_{\alpha\lambda\beta} A^{(1)}_{\lambda} A^{(2)}_{\beta}$$

Alternatively, since the superscripts are the same as column indices for the original direction cosine matrix,

(25a)
$$A_{\alpha 3} = \varepsilon_{\alpha\lambda\beta} A_{\lambda 1} A_{\beta 2}$$

Correspondingly,

(25b)
$$A_{\alpha 1} = \varepsilon_{\alpha\lambda\beta} A_{\lambda 2} A_{\beta 3}$$

(25c)
$$A_{\alpha 2} = \varepsilon_{\alpha\lambda\beta} A_{\lambda 3} A_{\beta 1}$$

These are extremely useful formulae, well-known since the Eighteenth Century. In earlier times they were gotten by working out the elements of the inverse matrix of A in a "brute force" manner.

Equations 25 can be used to find the determinant of a direction cosine matrix. Expanding the determinant following the first column,

$$\det A = A_{11}(A_{22}A_{33} - A_{23}A_{32}) - A_{21}(A_{12}A_{33} - A_{13}A_{32}) + $$
$$+ A_{31}(A_{12}A_{23} - A_{13}A_{22})$$

By Eq. 25b with $\alpha = 1$, the first parenthesized grouping is A_{11}; with $\alpha = 2$ and 3 the same equation gives the second and third groupings respectively as $-A_{21}$, A_{31}. Hence

$$\det A = (A_{11})^2 + (A_{21})^2 + (A_{31})^2 = 1 \tag{26}$$

Next consider the eigenvalue problem for a direction cosine matrix: to find values of λ such that

$$\det(A - \lambda E) = 0 \tag{27}$$

Expanding the equation into a polynomial in λ,

$$\lambda^3 - (\mathrm{tr}\ A)\lambda^2 + (\mathrm{tr}\ A)\lambda - 1 = 0 \tag{28a}$$

Here $\mathrm{tr}\ A = \mathrm{trace}\ A = A_{\alpha\alpha}$, Eqs. 25 have been used to reduce differences of products of the elements, and we have taken advantage of Eq. 26 in setting $\det A = 1$. It is evident that Eq. 28 can be rewritten as

$$(\lambda - 1)\left[\lambda^2 + (1 - \mathrm{tr}\ A)\lambda + 1\right] = 0 \tag{28b}$$

It follows that $\lambda = 1$ is an eigenvalue of A. To find the other two, let $\underline{u}$ be the unit eigenvector corresponding to the real eigenvalue 1. It is shown in Fig. 1. The rotation carries $\underline{e}_1$ into $\underline{e}_1'$, on each of which vectors $\underline{u}$ has the same component u_1 (this is the meaning of an eigenvector). Remember that u_1 is the cosine of the angle between $\underline{u}$ and $\underline{e}_1$ and between $\underline{u}$ and $\underline{e}_1'$. Inasmuch as $\underline{e}_1$, $\underline{e}_1'$ and $\underline{u}$ are unit vectors, their tips, geometrically speaking, lie on a surface of unit radius. On this surface, ABC is a spherical triangle of which the sides AB and AC are both known to be equal to arccos u_1. Furthermore, the cosine of the angle between $\underline{e}_1$ and $\underline{e}_1'$ is A_{11}. Finally, the dihedral angle Φ between the planes OAB and OAC is the rotation about the eigenvector to carry $\underline{e}_\alpha$ into $\underline{e}_\alpha'$ for all α. The law of cosines of spherical trigonometry gives

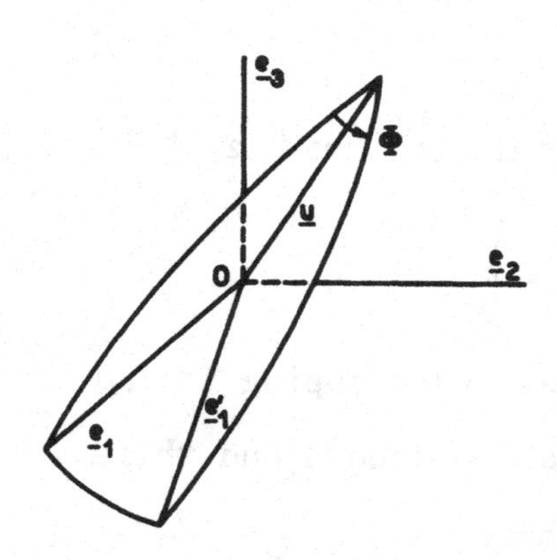

Fig. 1. Rotation about eigenvector

$$\cos \langle \underline{e}_1, \underline{e}_1' \rangle = \cos(\arccos u_1)\cos(\arccos u_1) + $$

$$+ \sin(\arccos u_1)\sin(\arccos u_1)\cos\Phi$$

or

$$A_{11} = (u_1)^2 + \left(\sqrt{1-(u_1^2)}\right)^2 \cos\Phi =$$

$$= \cos\Phi + (u_1)^2(1-\cos\Phi) \tag{29a}$$

By analogous argument,

$$A_{22} = \cos\Phi + (u_2)^2(1-\cos\Phi) \quad (29b)$$

$$A_{33} = \cos\Phi + (u_3)^2(1-\cos\Phi) \quad (29c)$$

Adding these,

$$\mathrm{tr}\ A = 3\cos\Phi + (1-\cos\Phi) = 1+2\cos\Phi \quad (30)$$

Returning to the quadratic factor in Eq. 28b, it can be rewritten

$$\lambda^2 + 2(\cos\Phi)\lambda + 1 = 0 \quad (31)$$

whose roots are complex, namely

$$\lambda = \exp(\pm i\Phi) \quad (32)$$

It follows that $\lambda = 1$ is the only real eigenvalue and the eigenvector $\underline{u}$ (Fig. 1) corresponding to it is the unique eigenvector.

Direction cosine matrices in chains of transformations compound by multiplication. Suppose that $\underline{e}^2 = A^{21}\underline{e}^1$ and that $\underline{e}^3 = A^{32}\underline{e}^2$. Then $\underline{e}^3 = A^{32}A^{21}\underline{e}^1$ and we can write

$$A^{31} = A^{32}A^{21} \quad (33)$$

Any number of successive rotations is handled the same way.

<u>Kinematics</u>

A rotation expressed by a direction cosine can

be described by three independent generalized coordinates, or by more if the equations of constraint among them is recognized. The most familiar parametrization is by a sequence of rotation angles about one coordinate axis at a time. Canonical rotation matrices are defined as

$$ {}^{1}A(\Theta) = \begin{bmatrix} 1 & 0 & 0 \\ 0 & \cos\Theta & \sin\Theta \\ 0 & -\sin\Theta & \cos\Theta \end{bmatrix} \tag{34a} $$

$$ {}^{2}A(\Theta) = \begin{bmatrix} \cos\Theta & 0 & -\sin\Theta \\ 0 & 1 & 0 \\ \sin\Theta & 0 & \cos\Theta \end{bmatrix} \tag{34b} $$

$$ {}^{3}A(\Theta) = \begin{bmatrix} \cos\Theta & \sin\Theta & 0 \\ -\sin\Theta & \cos\Theta & 0 \\ 0 & 0 & 1 \end{bmatrix} \tag{34b} $$

A rotation of frame e about axis $\underline{e}^{1}$ through angle φ to frame e^{2} is described by

$$ \underline{e}^{2} = {}^{\alpha}A(\varphi)\,\underline{e}^{1} \tag{35} $$

Thus $A^{21} = {}^{\alpha}A(\varphi)$ and $A^{12} = \left[{}^{\alpha}A(\varphi)\right]^{T} = {}^{\alpha}A(-\varphi)$. If the sequence were continued by rotating about $\underline{e}^{2}_{\beta}$ through angle Θ to frame e^{3},

$$ \underline{e}^{3} = {}^{\beta}A(\Theta)\,\underline{e}^{2} = {}^{\beta}A(\Theta)\,{}^{\alpha}A(\varphi)\,\underline{e}^{1} \tag{36} $$

(Here $\beta \neq \alpha$, or the rotation amounts to a single one through angle $\varphi + \Theta$)

In a three-axis rotation (through angle ψ , say), one has two basic choices.

1. The third axis label is the same as the first:

$$\underline{e}^4 = {}^{\alpha}A(\psi)^{\beta}A(\Theta)^{\alpha}A(\varphi)\underline{e}^1 \qquad (37a)$$

This choice is called an Euler sequence, and the angles are called Euler angles.

2. The third axis label γ is different from α and β :

$$\underline{e}^4 = {}^{\gamma}A(\psi)^{\beta}A(\Theta)^{\alpha}A(\varphi)\underline{e}^1 \qquad (37b)$$

For historical reasons, this choice is called a Tait-Bryan sequence, and the angles are called Tait-Bryan angles.

The "angular velocity"is defined as the instantan taneous rate of rotation about the instantaneous eigenvector of rotation.

$$\underline{\omega} = \lim_{\Delta t \to 0} (\Delta\Phi / \Delta t)\underline{u} \qquad (38)$$

where $\underline{u}$ is the eigenvector and Φ the rotation angle. Consider the rotation from frame e to frame $e'(t)$. We can regard the rotation from e to $e'(t + \Delta t)$ to be compounded of two rotations: that to $e'(t)$ followed by an infinitesimal rotation from $e'(t)$

to $e'(t+\Delta t)$. The latter is denoted

$$
(39) \quad \begin{aligned} {}^{u}A(\Delta\Phi) &= E - \tilde{u}\sin\Delta\Phi + \tilde{u}\tilde{u}(1-\cos\Delta\Phi) \\ &\approx E - \tilde{u}\,\Delta\Phi + \dots \end{aligned}
$$

(This is a standard form for a canonical rotation about an arbitrary axis $\underline{u}$.) Thus $A(t+\Delta t) = (E - \tilde{u}\,\Delta\Phi)A(t)$, from which it follows, in view of Eq. 38, that

$$
(40) \quad \frac{dA}{dt} = -\tilde{\omega} A
$$

or, if solved for $\tilde{\omega}$,

$$
(41) \quad \tilde{\omega} = A\dot{A}^{T}
$$

Given the change in the direction cosine matrix, the angular ve- is found from Eq. 41. Given the angular velocity, the change in orientation is found by solving the differential equation, Eq. 40. The latter, called here by the name "Euler-Lagrange-Poisson", can be considered the basic equation of kinematics.

The vector

$$
(42) \quad \omega_{\alpha} = \frac{1}{2}\,\epsilon_{\alpha\mu\lambda}\,\omega_{\lambda\mu}
$$

is the angular velocity of frame e' with respect to frame e , components resolved in e' . More generally we write $\underline{\omega}^{ij}$ of frame i with respect to frame j, derived from direction cosine matrix A^{ij} . In a single axis rotation from e to e' about axis $\underline{e}_{\alpha}$ through angle Θ , the vector angular velocity is

$$\underline{\omega} = \underline{e}_{\alpha}\dot{\theta} \tag{43}$$

Relative derivations are very important in the rotation of complex systems. Consider two frames e^2 and e^1 having a relative angular velocity $\underline{\omega}^{21}$ resolved in the $\underline{e}^2$-frame as ω^{21}. Denote by ${}^1\overset{\circ}{\underline{v}}$ the "relative derivative" in frame e^1 of an arbitrary vector $\underline{v}$: that is,

$${}^1\overset{\circ}{\underline{v}} = \underline{e}^{1^T} \frac{dv}{dt} \tag{44}$$

Derivatives are taken only of the vector components without regard for any possible rotation of the base vectors themselves. A basic formula is

$${}^1\overset{\circ}{\underline{v}} = {}^2\overset{\circ}{\underline{v}} + \underline{\omega}^{21} \times \underline{v} \tag{45}$$

<u>Dynamics</u>

Angular momentum with respect to a point O fixed in inertial space is defined as

$$\underline{H} = \int_{\mathfrak{M}} \underline{\rho} \times \frac{d\underline{\rho}}{dt}\, dm \tag{46}$$

Torque on the system about the same point from a field of applied specific forces $\underline{f}$ (forces per unit mass) is

$$\underline{L} = \int_{\mathfrak{M}} \underline{\rho} \times \underline{f}\, dm \tag{47}$$

To this is added any body torque not resulting from moments of force.

The basic law of dynamics is that

$$\frac{d\underline{H}}{dt} = \underline{L} \tag{48}$$

Remember: this form holds when O is a fixed point. From it, one can easily show that the same form of the dynamical equation holds when O is the center of mass of the system, but not for arbitrary base points.

The angular momentum can be split into two parts, and none of these factored into a part inherent to the body (the moment of inertia tensor) and the angular velocity. Let frame X be associated with the material system is a rigid body or gyrostat), and have angular velocity $\underline{\omega}$ in inertial space. Then Eq. 46 can be rewritten

$$\begin{aligned} \underline{H} &= \int_{\mathfrak{M}} \underline{\rho} \times (\overset{\circ}{\underline{\rho}} + \underline{\omega} \times \underline{\rho}) \, dm = \\ &= \underline{\rho} \times \overset{\circ}{\underline{\rho}} \, dm + \left(\int_{\mathfrak{M}} (\underline{\rho} \cdot \underline{\rho} \, E - \underline{\rho} \underline{\rho}) \, dm \right) \cdot \underline{\omega} \end{aligned} \tag{49}$$

The first of these integrals is called the "relative internal angular momentum" with respect to point O . Here we denote it by $\underline{h}$. In the gyrostat it is the relative angular momentum of spin of the internal rotor with respect to the rest of the body. The second integral is the "inertia dyadic" with respect to the point O. It is denoted here by **J** if the point O is a general fixed point and by **I** if it is the center of mass of the system.

Thus, in vector form,

$$\underline{H} = \mathbf{J} \cdot \underline{\omega} + \underline{h} \tag{50}$$

with $\underline{h}$ a constant vector as resolved in body axes, and $\mathbf{J}$ a constant dyadic as resolved in body axes, for a classical gyrostat. The relative derivative of $\underline{H}$ in body axes is just $\mathbf{J} \cdot \dot{\underline{\omega}}$, so the gyrostat dynamical equation in vector-dyadic form is

$$\mathbf{J} \cdot \dot{\underline{\omega}} + \underline{\omega} \times (\mathbf{J} \cdot \omega + \underline{h}) = \underline{L} \tag{51}$$

In matrix form it is

$$J \dot{\omega} + \tilde{\omega} (J \omega + h) = L \tag{52}$$

3. Angular Velocity of Free Rigid Body

The general problem is governed by the Euler rotational equations

$$J \dot{\omega} + \tilde{\omega} J \omega = 0 \tag{1}$$

in which the external torques L and internal angular momentum have been set equal to zero. It is assumed that one point of the body is not accelerating, or else that it is the center of mass, to permit the equation to be written in this form. The inertia matrix J is appropriate to that point of the body. Body axes $\{\underline{X}_\alpha\}$ are assumed to be principal axes of inertia, so J contains only the diagonal elements (principal moments of inertia) J_α. If body axes are labeled arbitrarily, there is no a priori ordering relationship among these J_α. To get a canonical form of the solution it is assumed without loss of generality that the body axis $\underline{X}_2$ is the axis of intermediate moment of inertia. Furthermore, when we define the location of body axes with respect to inertial axes by an Euler angle sequence, we use the sequence 3,1,3 so the axis $\underline{X}_3$ is the axis to be regarded as the "spin axis". We denote the three principal moments of inertia by A, B, C and distinguish two cases:

(2)
$$A > B > C \qquad \text{Prolate}$$
$$A < B < C \qquad \text{Oblate}$$

(The "prolateness" or "oblateness" is with reference to the spin axis, $\underline{X}_3$.) In these terms the scalar forms of Eq. 1 are

(3a)
$$A\dot{\omega}_1 + (C-B)\,\omega_2\,\omega_3 = 0$$

(3b)
$$B\dot{\omega}_2 + (A-C)\,\omega_3\,\omega_1 = 0$$

(3c)
$$C\dot{\omega}_3 + (B-A)\,\omega_1\,\omega_2 = 0$$

We first get two integrals of these dynamical equations. Multiply Eqs. 3 by $\omega_1, \omega_2, \omega_3$ respectively and add, then integrate the result to get

(4a)
$$A(\omega_1)^2 + B(\omega_2)^2 + C(\omega_3)^2 = 2T$$

The constant of integration $2T$ is twice the kinetic energy of rotation and Eq. 4a is the energy integral of the system. In an analogous way, multiply Eqs. 3 by $(A\omega_1)(B\omega_2)$, $(C\omega_3)$, add and integrate. The resulting second integral of the Euler equations is

$$(A\omega_1)^2 + (B\omega_2)^2 + C(\omega_3)^2 = H_0^2 \tag{4b}$$

in which the constant H_0 is the magnitude of the body's angular momentum. There are no algebraic integrals of Eqs. 3 other than Eqs. 4. It has been customary since the time of Poisson to introduce a parameter D having the dimensions of moment of inertia, defined by

$$D = H_0^2 / 2T \tag{5}$$

and used in Eq. 4b to replace H_0^2. The integrals describe two ellipsoids in the space whose rectangular cartesian coordinates are ω_α, called respectively the energy ellipsoid and angular momentum ellipsoid. Their intersection represents the locus of the angular velocity vector as the motion proceeds.

Consider the projections of this locus into the coordinate planes of the ω_α coordinate system. The three projections can be gotten by eliminating $\omega_1, \omega_2, \omega_3$ one at a time from Eqs. 4, with the three results

$$B(A-B)\omega_2^2 + C(A-C)\omega_3^2 = 2T(A-D) \tag{6a}$$

(6b) $$A(A-B)\omega_1^2 + C(C-B)\omega_3^2 = 2T(D-B)$$

(6c) $$A(A-C)\omega_1^2 + B(B-C)\omega_2^2 = 2T(D-C)$$

From Eqs. 6a and 6c emerges a physical range of the parameter D. If $A > C$, their left sides are obviously positive, so $A \geq D \geq C$. Conversely, if $A < C$ their left sides are negative and $C \geq D \geq A$. Hence

(7) $$\min(A,C) \leq D \leq \max(A,C)$$

Furthermore, Eq. 6 gives a good geometrical picture of the angular velocity loci. Writing it as

(8a) $$\frac{A(A-B)}{(D-B)}\omega_1^2 - \frac{C(B-C)}{(D-B)}\omega_3^2 = 1$$

it becomes obvious that the $\underline{X}_1$, $\underline{X}_3$ -projections are hyperbolae for all values of $D \neq B$. For $D = B$ one obtains the hyperbolic asymptotes

(8b) $$\omega_3 = \pm\sqrt{\frac{A(A-B)}{C(B-C)}}\,\omega_1$$

(whether $A > C$ or $C > A$). If $D > B$ and $A > B$, or if $D < B$ and $A < B$, there are real intersections of the hyperbolae on the ω_1-axis.

If $D < B$ and $A > B$ or $D > B$ and $A < B$, there are real intersections on the ω_3-axis. The projections of typical loci in the ω_1, ω_3-plane are shown in Fig. 1 for $A > C$. To develop a complete canonical solution we have to consider two solutions cases:

$$(D-B)(A-B) > 0 \qquad \underline{\text{Sidecap}} \text{ solution}$$
$$(D-B)(A-B) < 0 \qquad \underline{\text{Endcap}} \text{ solution}$$

together with the separating condition $D = B$. In Fig. 1 the arrow-heads show the direction the angular velocity vector traverses its locus with increasing time. Equation 3b, in particular, shows that $\operatorname{sgn} \dot{\omega}_2 = \operatorname{sgn}\left[(C-A)\omega_1\omega_3\right]$, which tells us for each quadrant of the figure whether ω_2 is increasing (locus passing from visible toward invisible side ellipsoid, i.e. away from the reader) or decreasing (locus coming toward the reader).

Another geometrical viewpoint could have been used to arrive to Eq. 7. It is not really needed in the present problem, but is noteworthy because it represents a route that can be followed successfully for the gyrostat (discussed in the next lectures) when the simpler route does not offer itself. The alternative can be sufficiently small, the ellipsoid represented by Eq. 4b would be entirely within the energy ellip-

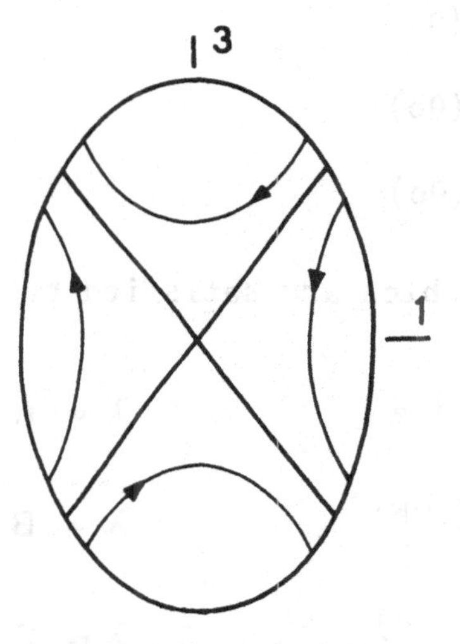

Fig. 1. Polhodes on the Energy Ellipsoid

soid (Eq. 4a) and the two would not intersect. As D increases the ellipsoids finally osculate,then intersect in a curve. Analogously, if D were very large the energy ellipsoid would be entirely within the momentum ellipsoid: as D decreases, the two first osculate and they intersect in a curve. The physically meaningful range of D, therefore, is the range between the minimum and maximum D-values which cause osculation. Now osculation occurs when the gradients of the two ellipsoid are parallel at the intersection point(s). That is, when

$$2A^2\omega_1\underline{X}_1 + 2B^2\omega_2\underline{X}_2 + 2C^2\omega_3\underline{X}_3 =$$

$$= \lambda(2A\omega_1\underline{X}_1 + 2B\omega_2\underline{X}_2 + 2C\omega_3 X_3)$$

where λ is some constant of proportionality. This is equivalent to the three equations

(9a) $$(A - \lambda)\omega_1 = 0$$

(9b) $$(B - \lambda)\omega_2 = 0$$

(9c) $$(C - \lambda)\omega_3 = 0$$

which are satisfied by

(10a) $$\lambda = A, \quad \omega^T = \left[\pm\sqrt{2T/A} \quad 0 \quad 0\right]$$

(10b) $$\lambda = B, \quad \omega^T = \left[0 \quad \pm\sqrt{2T/A} \quad 0\right]$$

(10c) $$\lambda = C, \quad \omega^T = \left[0 \quad 0 \quad \pm\sqrt{2T/A}\right]$$

Using these ω-components in Eq. 4b in turn, we get respectively $D = A$, $D = B$, $D = C$. Clearly the range of D is established by the extreme values A and C, whence its physical range is as given by Eq. 7.

To develop a formal solution, we use Eq. 6 to get ω_1 and ω_3 as functions of ω_2, and substitute the results into Eq. 3b. We get

$$\omega_1 = \pm\sqrt{\left[2T(D-C) - B(B-C)\omega_2^2\right]/A(A-C)} \tag{11a}$$

$$\omega_3 = \pm\sqrt{\left[2T(A-D) - B(A-B)\omega_2^2\right]/C(A-C)}$$

$$\dot{\omega}_2 = \epsilon\sqrt{1/AB^2C}\sqrt{\left[2T(D-C) - B(B-C)\omega_2^2\right]\left[2T(A-D) - B(A-B)\omega_2^2\right]} \tag{12}$$

where $\epsilon = \pm 1$. The sign ambiguity is resolved as described above, leading to

$$\epsilon = \operatorname{sgn}\left[(C-A)\omega_{10}\omega_{30}\right] \tag{13}$$

where we take as intial conditions for the canonical solution $\omega_1 = \omega_{10}$, $\omega_2 = 0$, $\omega_3 = \omega_{30}$. Inasmuch as <u>all</u> loci pass through the plane $\omega_2 = 0$, this choice assures that the standard form incompasses all possible situations as ω_{10} and ω_{30} respectively range over $\pm\sqrt{2T/A}$ and $\pm\sqrt{2T/C}$, subject to the constraint that these initial conditions satisfy the energy integral. (We regard $2T$ as a fixed constant during the investigation.)

Now define

$$a^2 = \max \atop b^2 = \min \Bigg\} \left(\frac{2T(D-C)}{B(B-C)}, \frac{2T(A-D)}{B(A-B)} \right) \tag{14}$$

It is easy to show that for the endcap solution

$$b^2 = \frac{2T(D-C)}{B(B-C)} < \frac{2T(A-D)}{B(A-B)} = a^2 \tag{15a}$$

whereas for the sidecap solution

$$a^2 = \frac{2T(D-C)}{B(B-C)} > \frac{2T(A-D)}{B(A-B)} = b^2 \tag{15b}$$

One now can write the integral of Eq. 12 as

$$\int_0^{\omega_2} \frac{dx}{\sqrt{(a^2-x^2)(b^2-x^2)}} = \epsilon \sqrt{\frac{(A-B)(B-C)}{AC}}\, t \tag{16}$$

The integral is expressed in terms of inverse elliptic functions by any standard table. Define modulus and frequency as

$$k = b/a \tag{17}$$

$$p = \epsilon a \sqrt{\frac{(A-B)(B-C)}{AC}} \tag{18}$$

(In all cases, a and b are to be interpreted as the positive square roots of a^2 and b^2.) Then

$$\omega_2 = b\,\mathrm{sn}(pt, k) \tag{19}$$

Hereafter the modulus of the elliptic function is not written unless required for clarity.

It is easy enough to get ω_1 and ω_3 by returning with Eq. 19 to Eqs. 11, distinguishing the two solution regimes. It is found that

$$\omega_1 = \pm \frac{B(B-C)}{A(A-C)} \times \begin{cases} b \ \mathrm{cn}\ pt & \text{(end cap)} \\ a \ \mathrm{dn}\ pt & \text{(side cap)} \end{cases} \tag{20a}$$

$$\omega_3 = \pm \frac{B(A-B)}{C(A-C)} \times \begin{cases} a \ \mathrm{dn}\ pt & \text{(end cap)} \\ b \ \mathrm{cn}\ pt & \text{(side cap)} \end{cases} \tag{20b}$$

At this point the angular velocity is completely known through Eqs. 19 and 20, albeit with some sign ambiguity in the latter. However, the coefficients are considerably more complicated than necessary, referring as they do to the initial conditions only through the indirect medium of parameter D, which then is buried further in parameters a and b. We can simplify the coefficients and resolve the sign ambiguity at the same time if we cut to the heart of the matter, and note that initial values of $\mathrm{cn}\ pt$ and $\mathrm{dn}\ pt$ are both unity. Thus the complete coefficients before these functions must represent the initial values ω_{10} or ω_{30}, as the case may be. In summary, we can write

$$\omega_1 = \omega_{10}\Omega_1(t) \tag{21}$$

$$\omega_3 = \omega_{30}\Omega_3(t) \tag{22}$$

with Ω_1 and Ω_3 given by Table 1. An analogous simplification can be made in the case of ω_2, but the rationale is different. It depends on the fact that the initial value of $\dot{\omega}_2$ must satisfy Eq. 3b. Thus we are compelled to choose ω_2 in the form

$$\omega_2 = \frac{(C-A)\omega_{10}\omega_{30}}{Bp}\,\mathrm{sn}\,pt \tag{23}$$

Table 1 Functions Ω_1 and Ω_3

Case	Ω_1	Ω_3
End cap, $(D-B)(A-B) < 0$	cn pt	dn pt
Side cap, $(D-B)(A-B) > 0$	dn pt	cn pt

It should be noted how the solutions behave for bodies symmetric about the spin axis, so $A=B$. When $A=B$ the side cap conditions cannot be fulfilled, so only the end cap solution is of concern. Equation 15a implies $a=\infty$, whence $k=0$; but $a\sqrt{A-B} = \sqrt{2T(A-B)/B}$, whence it can be found that $p = -(\mathrm{sgn}\,\omega_{10})(A-C)\omega_{30}/A$. The forms in which Eqs. 21, 22 and 23 are written present no difficulty when $A=B$, for they do not explicitly involve the difference.

4. The Free Gyrostat

The gyrostat is a relatively obvious and stright-forward generalization of the rigid body, although the problem of a gyrostat spinning about a fixed point in the absence of external torque is a century junior to the corresponding problem of the free rigid body. The pertinent dynamical equation is

$$I\dot{\omega} + \tilde{\omega}(I\omega + h) = 0 \tag{1}$$

with I and h constant matrices. This is written as if the centre of mass were the fixed point, but if any other point were the fixed point instead, the I-matrix would simply be replaced by the proper J-matrix.

If Eq. 1 be premultiplied successively by ω^T and $(I\omega + h)^T$ and the two results integrated, we find that two integrals of the motion are

$$\omega^T I \omega = 2T \qquad\qquad (I\omega + h)^T (I\omega + h) = H_0^2 \tag{2}$$

Here T has the physical interpretation of kinetic energy of a rigid body rotating like the gyrostat; that is, as if the kinetic energy embodied in the relative rotation of the rotor were not counted. The constant H_0 is the magnitude of the total angular momentum of the system, counting both the gyrostat body and the rotor, the integrals are used in the sequel for the case where principal axes of inertia are used as body axes, so Eqs. 2 can be rewritten

$$f_1 = (I_1\omega_1 + h_1)^2 + (I_2\omega_2 + h_2)^2 + (I_3\omega_3 + h_3)^2 - H_0^2 = 0$$

$$f_2 = I_1\omega_1^2 + I_2\omega_2^2 + I_3\omega_3^2 - 2T = 0$$

Inspection of Eqs. 3 quickly reveals the added complications from the internal rotor. First, the explicit determination of the D-range giving real polhodes is considerably more difficult, in general, for the momentum ellipsoid is not centered at the same point as the osculation points depend not only on D but also on all three of the components of $\underline{h}$. Second, when the true D-value is given in a physical problem, one cannot generally solve Eqs. 3 explicitly for any of the angular velocity components in terms of one of the others, thus is prevented from following precisely the same path as in the rigid body case. Nevertheless, procedures do exist for obtaining equations for the polhodes in terms of a new auxiliary variable, and for relating the angular velocity components to time through the dynamical equations. Carrying them out in practice, however, can be accomplished in literal terms only in special cases. In more general cases, the procedure first requires the solution of a certain sextic polynomial equation which only can be done numerically.

Before proceeding further, it is desirable to establish that two classical problems, superficially somewhat different, both have the same dynamical description as the gyrostat in the sense the word has been used heretofore. This means that all the literature of both problems actually are applicable to the analysis undertaken later in this section. We follow the notational conventions for two rigid bodies joined at a hinge point fixed in both bodies. Visualize a hinge axis $\underline{u}$ fixed in both bodies and passing through this point. Suppose that the center of mass of body 2 is on this axis, and that body 2 is symmetric about the axis. Let K^1 be the inertia matrix of the two-body system about its center of mass, I^2 be that of the rotor alone. External torques on both are zero. $\underline{\mathcal{L}}^{12}$ is interaction torque on main body by rotor, and $\underline{\mathcal{L}}^{21}$ is its negative. Define:

<u>Problem 1</u>. Body 2 is driven with respect to body 1 at constant angular velocity about the common axis;

<u>Problem 2</u>. The interaction torque of body 1 on body 2 has a zero component on the common axis.

The first gives the gyrostat of Eq. 1a, the second gives a system originally stidied by Wangerin (1889), later by Leipholz (1962).

We now show that the second problem leads to a dynamical equation having exactly the same form as for the first. Applying Eq. 2.52 separately to the system and to the rotor,

$$K^1\dot{\omega}^1 + \tilde{\omega}^1 K^1\omega^1 = \left[\underline{\mathcal{L}}^{12}\cdot \underline{X}^1_\alpha\right] \tag{4a}$$

$$I^2\dot{\omega}^2 + \tilde{\omega}^2 I^2\omega^2 = \left[\underline{\mathcal{L}}^{21}\cdot X^2_\alpha\right] \tag{4b}$$

These are accompanied by the kinematical relations

$$\omega^2 = A^{21}(\omega^1 + \dot{\sigma} u) \tag{5a}$$

$$\dot{\omega}^2 = A^{21}(\dot{\omega}^1 + \ddot{\sigma} u + \dot{\sigma}\tilde{\omega}^1 u) \tag{5b}$$

where $\dot{\sigma}$ is the spin speed of body 2 with respect to body 1 about the axis and u is the resolution of the axis $\underline{u}$ in the body 1 - frame. Multiplying Eq. 4b by A^{21}, and using Eqs. 5 together with the fact that $A^{12} I^2 A^{21} = I'$ is simply the inertia matrix of body 2 resolved in the body 1 frame - constant by virtue of the symmetry of body 2 about the axis of rotation - Eq. 4b reduces to

$$I'\dot{\omega}^1 + \tilde{\omega}^1 I'\omega^1 + \ddot{\sigma} I' u + \dot{\sigma}(I'\tilde{\omega}^1 u + \tilde{u} I'\omega^1 + \tilde{\omega}^1 I' u) = \\ = [\underline{\mathcal{L}}^{21} \cdot \underline{X}^1_\alpha] \tag{6}$$

The fact that the eigenvector is an axis of symmetry of the body implies that the inertia matrix can be written in the form $I' = I_0 E + (\lambda - I_0) u u^T$, where I_0 is the moment of inertia about the axes normal to the eigenvector u and λ is the polar moment of inertia about the eigenvector. In turn, this implies both that $I'u = \lambda u$ and that $I'\tilde{\omega}^1 u + \tilde{u} I'\omega^1 = (\tilde{u} I' - I'\tilde{u})\omega^1 = 0$. Defining $h = \lambda \dot{\sigma} u$, Eq. 6 reduces to

$$I'\dot{\omega}^1 + \tilde{\omega}^1 I'\omega^1 + \dot{h} + \tilde{\omega}^1 H = [\underline{\mathcal{L}}^{21} \cdot \underline{X}^1_\alpha] \tag{7}$$

If Eq. 7 be added to Eq. 4a and we define

$$K = K^1 + I' \tag{8}$$

then we get

$$K\dot{\omega}^1 + \tilde{\omega}^1(K\omega^1 + h) + \dot{h} = 0 \tag{9}$$

In Problem 1, of course, $\dot{h} = 0$ and we immediately get the equation for the classical gyrostat. In this case, furthermore, the center of mass of the augmented body 1 coincides with the system center of mass, and K is easily interpretable as the inertia matrix of the composite system about its center of mass, normally denoted I in this work. In Problem 2, we premultiply Eq. 7 by u^T to obtain

$$u^T I'\dot{\omega}^1 + u^T\tilde{\omega}^1 I'\omega^1 + u^T h + u^T\tilde{\omega} h = u^T\left[\mathcal{L}^{21}\cdot \underline{X}^1_\alpha\right] = \underline{u}\cdot\mathcal{L}^{21} = 0$$

The last equality follows from the postulate that the interaction torque has a zero component on the common axis. Moreover, because h is parallel to $\underline{u}$, $u^T\tilde{\omega}^1 h = -u^T\tilde{h}\omega^1 = 0$. Finally, we can regroup the second term in the equation as $u^T\tilde{\omega}^1 I'\omega^1 = -\omega^{1T}\tilde{u}I'\omega^1 = -\omega^{1T}I'\tilde{u}\omega^1 = \omega^{1T}I'\tilde{\omega}^1 u = -(u\tilde{\omega}^1 I'\omega^1)^T$ and conclude that the term is zero because it is a scalar equal to to the negative of its transpose. Thus we are led to

$$u^T(I'\dot{\omega}^1 + \dot{h}) = 0 \tag{10}$$

This can be integrated to give an angular momentum integral for the second body, namely

$$u^T(I'\omega^1 + h) = \eta \quad \text{(constant)} \tag{11}$$

It remains to premultiply Eq. 10 by u, Eq. 11by $\tilde{\omega}^1 u$, and subtract both results from Eq. 9, noting that $uu^T\dot{h} = u\lambda\ddot{\sigma} = \dot{h}$ and $u^T I' = \lambda u^T$, to get

$$(12) \qquad (K - \lambda uu^T)\dot{\omega}^1 + \tilde{\omega}^1\left[(K - \lambda uu^T)\omega^1 + \eta u\right] = 0$$

Evidently, this has exactly the same structure as the gyrostat equation provided parameters are used per Table 1. This type of two-body system satisfying the conditions of Problem 2 has been named an apparent gyrostat.

Table 1. Identification of Canonical Parameters in the Equation

Case	I	h
Gyrostat	$K^1 + I^1$	$(\lambda\dot{\sigma})u$
Apparent gyrostat	$K^1 + I' - \lambda uu^T$	ηu

Recall Euler's problem for a rigid body treated in § 3 What are the conditions under which a body can turn freely about an axis? The same question may be asked about the gyrostat. It was considered briefly by Volterra (1898), but his solution was not complete in certain respects: it was phrases only in terms of the "points multiples de la polhodie", which themselves were not given incisive analytical characterization. We can anticipate that the axis must skewer the center of mass as in the case of a rigid body, so we focus on the rotational aspects of the problem <u>ab initio</u> by assuming that the center of mass is a space-fixed point through which passes the axis about which the body is free to turn. A necessary condition for it to turn "freely" is that its angular velocity about the axis, hence the body components of angular velocity, shall be constant in the absence of external torques; this condition also is sufficient. If ω be set constant in the dynamical equations, one finds that

$$\tilde{\omega}(I\omega + h) = 0 \tag{13}$$

is a necessary and sufficient condition for the solution of the generalized Euler problem.

There are two ways Eq. 13 can be satisfied. Either

$$I\omega + h = 0 \tag{14a}$$

or

$$I\omega + h = \lambda\omega \quad (\lambda \neq 0) \tag{14b}$$

The solutions of Eqs. 14ab are respectively denoted ω^+ and ω^* below. If Eq. 14a holds, we speak of the "zero-momentum gyrostat", one in which the internal angular momentum has three components that exactly negate the threee components of $I\omega$, where ω is regarded as given. Any axis of a zero-momentum gyrostat can be made a "permanent axis of rotation" (Euler's name for the axis that solves his problem) by choosing the internal angular momentum properly. Conversely, given h, the permanent axis is along $\omega^{+} = -I^{-1}h$. This axis direction is an inherent characteristic of the body, determined by I and h alone, so it would seem to represent a solution to Euler's problem in the sense that he intended for the rigid body. It is not quite the same, however, for the spin about that axis can take place only at a specific spin speed if the zero-momentum condition is not to be violated. In the original problem, the permanent axis was sought as a property of the body independent of the spin speed.

If Eq. 14a does not hold, then Eq. 14b becomes the condition to investigate. The vector ω^{*} satisfies

$$(I - \lambda E)\omega^{*} = -h \tag{15}$$

Because the right hand side is not zero, this is not an eigenvalue problem. If λ is a value such that the inverse matrix $(I - \lambda E)^{-1}$ exists, then $\omega^{*} = -(I - \lambda E)^{-1}h$. Using Eq. 15 in Eq. 3b, assuming as at the beginning of the section that body axes are principal axes of inertia,

$$\sum_{\alpha=1}^{3} \frac{I_{\alpha} h_{\alpha}^{2}}{(\lambda - I_{\alpha})^{2}} = 2T \tag{16}$$

If the I_{α} are all distinct and all $h_{\alpha} \neq 0$, Eq. 16 is equivalent to the sextic algebraic equation

$$\Sigma I_\alpha h_\alpha^2 (\lambda - I_\beta)^2 (\lambda - I_\gamma)^2 = 2T(\lambda - I_1)^2 (\lambda - I_2)^2 (\lambda - I_3)^2 \quad (17)$$

where α, β, γ are cyclic and the same summation range as in Eq. 16 is assumed implicitly hereafter, unless otherwise specified. If λ_0 now be any real root of Eq. 17, Eq. 15 gives

$$\omega^*_\alpha = -h_\alpha / (I_\alpha - \lambda_0) \qquad (\alpha = 1, 2, 3) \quad (18)$$

Substituting this result into Eq. 3a, one gets what we will call the <u>osculation D-value</u>, D_0, corresponding to λ_0, as

$$2TD_0 = \lambda_0^2 \Sigma \frac{h_\alpha^2}{(I_\alpha - \lambda_0)^2} \quad (19)$$

The solution of Euler's problem is easily shown to be the same as the determination of the two ellipsoids described by Eqs. 3. Setting the gradient of the surface $f_1 = 0$ proportional to the gradient of the surface $f_2 = 0$, we have the three equations

$$I_\alpha \omega^*_\alpha + h_\alpha = \lambda \omega^*_\alpha \quad \text{(not summed)} \quad (20)$$

This result is precisely the scalar form of the matrix Eq. 15.

The details of establishing the ω^*-matrices corresponding to osculation of the ellipsoids need to be discussed separately for the various h-cases and body symmetries. We observe in passing that there must be two, four or six such osculation vectors because the sextic algebraic equation, Eq. 17, must have an even number of real roots. We first dispose of the degenerate situations caused by the coalescence of two or more principal moments of inertia.

Case I ("sphere")

Suppose first that $I_1 = I_2 = I_3$. There is no loss in generality in supposing that the body axes are selected such that $h_1 = h_2 = 0$ and $h_3 > 0$. (We do not accept $h_3 = 0$, for this would return us to the case of a rigid body, already treated.) It is not hard to see in this case that the only solution of Eqs. 15 (or 20) is

$$\omega_1^* = 0\,, \quad \omega_2^* = 0\,, \quad \omega_3^* = h_3/(\lambda - I_3) \tag{21a}$$

Because the solution also satisfies the ellipsoid equations, Eqs. 3, it follows that

$$\omega_3^* = \pm\sqrt{2T/I_3}\,, \qquad (I_3\omega_3^* + h_3)^2 = 2TD \tag{21b}$$

From these, values of λ and D can be found if desired, but the central result is that the two osculation vectors are

$$\omega^* = \begin{bmatrix} 0 & 0 & \pm\sqrt{2T/I_3} \end{bmatrix}^T. \tag{22}$$

Case II ("cylinder")

Suppose next that exactly two principal moments of inertia are equal. There is no loss in generality in disposing the body axes such that $\underline{X}_3$ is normal to the plane of the equal principal moments: thus $I_1 = I_2 \neq I_3$. We consider two special subcases.

subcase II T ("transverse")

If $h_3 = 0$, one solution of Eq. 15 is

$$\left.\begin{aligned} \lambda = I_3 \ , \quad \omega_1^* = h_1/(I_3 - I_1) \ , \quad \omega_2^* = h_2/(I_3 - I_1) \\ \omega_3^* = \pm\sqrt{2T - I_1\omega_1^{*2} - I_2\omega_2^{*2}} \,/\, I_3 \end{aligned}\right\} \tag{23}$$

Another solution is gotten by finding $\lambda = I_1 \pm \sqrt{I_1 h_0^2/2T}$ from Eq. 16, whence Eq. 15 gives

$$\omega_3^* = 0 \ , \ \omega_1^* = \pm h_1/\sqrt{I_1 h_0^2/2T} \ , \ \omega_2^* = \pm h_2/\sqrt{I_2 h_0^2/2T} \tag{24}$$

where h_0 is the magnitude of $\underline{h}$ and the signs hold in the same order.

subcase II A ("axial")

If $h_1 = h_2 = 0$, one solution of Eq. 15 is again given by Eq. 22. Another is

$$\lambda = I_1 \ , \quad \omega_3^* = h_3/(I_3 - I_1) \tag{25a}$$

with ω_1^* and ω_2^* related by Eq. 3b as

$$\omega_1^{*2} + \omega_2^{*2} = (2T - I_3\omega_3^{*2})/I_1 \tag{25b}$$

This represents a whole plane of osculation vectors normal to the cylinder axis.

The nature of the special solutions for the sphere and for the axial and transverse cases of the cylinder are sketched in Fig. 1. The remaining possibility for the cylinder is that all components of internal angular momentum differ from zero.

subcase II P ("planar")

The left side of Eq. 16 is sketched in Fig. 2. If $T < T_0$, a certain critical value, there are exactly two real roots, whereas if the inequality be reversed there are four. The separating value is gotten easily by equating the derivative of the left side of Eq. 16 to zero. The minimum of the branch of the curve between $\min(I_1, I_3)$ and $\max(I_1, I_3)$ occurs at

$$\lambda = (I_3 k_1 + I_1 k_3) / (k_1 + k_3) \tag{26}$$

where

$$k_1 = \sqrt{I_1 (h_1^2 + h_2^2)} \qquad k_3 = \sqrt{I_3 h_3^2} \tag{27}$$

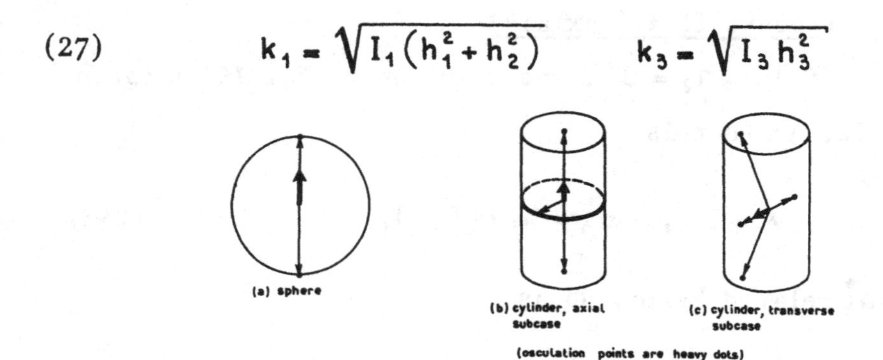

Fig. 1. Osculation vectors for Spherical and Cylindrical Bodies (*)

(*) Osculation points are heavy dots.

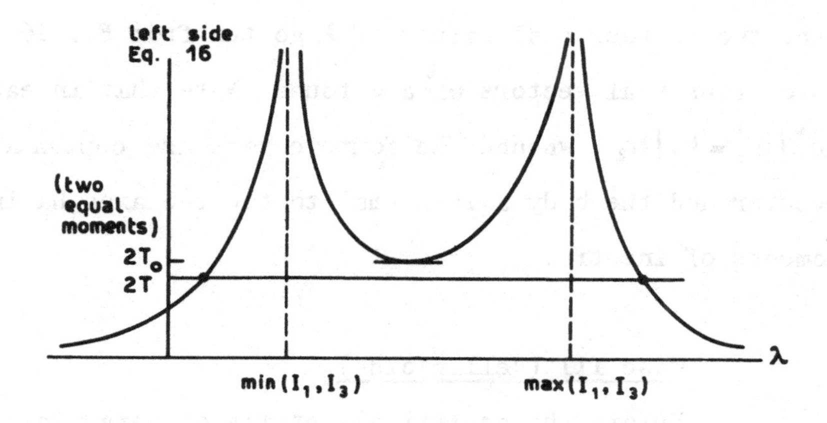

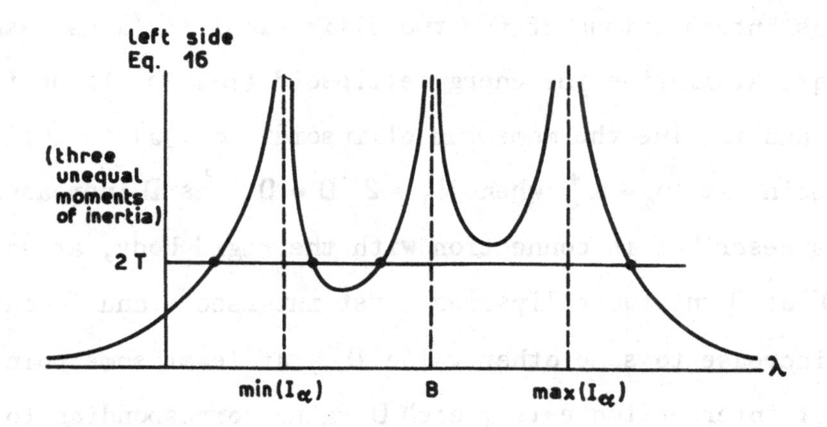

Fig. 2. Diagrams for Solutions of Eq. 16

Hence

$$2T_0 = (k_1 + k_3)^3 / (I_1 - I_3)^2 \tag{28}$$

Using the two or four real values of λ gotten from Eq. 16 in Eq. 15, two or four real vectors ω^* are found. Note that in each case $\omega_1^* / \omega_2^* = h_1 / h_2$, whence the four vectors are coplanar with the h-vector and the body axis normal to the two axes having equal moments of inertia.

Case III ("ellipsoid")

Before the analytical details of determining osculation vectors in this general case of the ellipsoidal gyrostat, let us visualize qualitatively the kinds of behavior that should be expected. To do this, we use the picture of the osculation points as intersections of the two ellipsoidal surfaces described by Eqs. 3. Imagine the energy ellipsoid (Eq. 3b) to be fixed in size and imagine the momentum ellipsoid (Eq. 3a) to swell up from a point at $\omega_\alpha = \omega_\alpha^+$ when $H_0^2 = 2TD = 0$, as D increases from zero. As described in connection with the rigid body, at some value D^- of D the two ellipsoids first intersect, and D continues to increase to some other value D^+, at least some points or curves of intersection exist, each D-value corresponding to a locus of angular velocity vectors in body axes. Finally, when $D > D^+$, the ellipsoids no longer intersect. For the rigid body, D^- is the minimum and D^+ the maximum moment of inertia. Part of the problem is to determine those extreme those extreme values of D between which all polhodes are generated, as functions of the magnitude and orientation of the internal angular momentum vector.

At least two real roots of Eq. 16 exist, whence there are at least two real osculation points. For if the left side of Eq. 16 be plotted as a function of λ , it has the general character shown in Fig. 2, the infinite singularities of function occurring when λ equals the principal moments of inertia. These principal moments are shown as distinct in the figure, as is appropriate to the present ellipsoidal case; also, the figure is drawn as if all components of internal angular momentum are non-zero, but this is not assumed in the text below. When the left side of the equation equals $2T$, Eq. 16 is satisfied. Four solution points are shown in the example sketched (*), but because the left side vanishes for $|\lambda|\to\infty$ it is evident that there are at least two solutions, no matter how small $2T$. One of these corresponds to $\lambda > \max(I_\alpha)$, the other to $\lambda < \min(I_\alpha)$. It also is evident that if, for fixed h_α, $2T$ is sufficiently large, there necessarily are six real solutions. Denote by $2T_1$ the transition level between two and four solutions, by $2T_2$ the transition level between four and six solutions. Thus $2T_1$ is the height of the bot-

(*) No inference is to be drawn that the two central solutions necessarily occur by intersections of the left cup of the function.

tom of the lower cup in Fig. 2, while $2T_2$ is the height of the bottom of the upper cup. These levels are analogues of the $2T_0$ given by Eq. 28 for the special case $I_1 = I_2$, but unfortunately they do not yield to so simple a representation.

It is important to recognize that the change of behavior with increasing $2T$ also can be regarded as a change with decreasing h_0 for fixed $2T$.

Suppose there are exactly two real solutions, λ_1 and λ_2, of Eq. 16. Equation 19 gives the corresponding osculation D-values, the smaller of which is denoted $D^{(1)}$ and the larger of which is denoted $D^{(2)}$. That is, $D^- = D^{(1)} < D^{(2)} = D^+$. As D is increases through this range, a progression of ellipsoid intersections is created as depicted in Fig. 3; D equal to one of the extreme values gives the single points where the osculation vectors pierce the energy ellipsoid, while other values give closed curves nested around the two osculation points.
If there are four real solutions λ_i of Eq. 16, there are four d-values form Eq. 19. Let these be ordered as

$$D^- = D^{(1)} < D^{(2)} < D^{(3)} < D^{(4)} = D^+ \tag{29a}$$

Any one osculation point must be either a local center or a local saddle point for polhode family: the topological requirements on the continuum of nested polhodes implies that only two possiblilities exists for the sequence $D^{(i)}$ when the osculation values are distinct, namely (center, center, saddle, center) or (center, saddle, center, center). One can easily convince himself that the first and last osculations cannot be saddles, and that two saddles cannot occur if there are only four osculation points. For the first of these sequences, closed curves appear about the osculation point for $D^- < D < D^{(2)}$. At $D = D^{(2)}$ another osculation point appears and ellipsoid intersection curves begin to grow around it. Finally, when $D = D^{(3)}$, the two families of curves touch in the third osculation point, which thus becomes a saddle point, and the complete curve through that osculation point (a "figure 8" wrapped around the ellipsoid) becomes a <u>separatrix</u>, or singular curve, of the families of polhodes on the ellipsoid. As D continues to increase, a nested family of closed curves is generated, containing the fourth osculation point (located on the far side of the ellipsoid). The only modification that can occur in the topology of the polhodes comes form the coalescence of λ-values in a double root of Eq. 16. The equation is sketched in Fig. 4, where the separation is a heavy line.

We turn to Fig. 5 for a schematic depiction of the polhodes when there six real (distinct) roots and six corresponding ordered D-values,

$$D^- = D^{(1)} < D^{(2)} < D^{(3)} < D^{(4)} < D^{(5)} < D^{(6)} = D^+ . \quad (29b)$$

In this case, if $D^{(3)} \neq D^{(4)}$, there are two "figure 8" separatrices, one wrapped around the two sides of the energy ellipsoid the other around the two ends.

Although we have examined the osculation problem in some detail, we still have not in so many words given the solution to Euler's problem. Reviewing the special osculation vectors obtained above, we find that the one described by Eq. 23 has a direction that depends on the rotational kinetic energy. In all other special cases, the direction of the osculation vector is a property of the body itself, with its internal rotor, thus represents a solution to the problem of "permanent axes" of the type we can suppose Euler had in mind for the rigid body. In the case of general body shape, suppose that the direction of ω^* is a constant, independent of $2T$. Then $\omega_1^*/\omega_3^* = k_1$ and $\omega_2^*/\omega_3^* = k_2$, where k_1 and k_2 are some constants. This implies that $h_i/(\lambda - I_i) = k_i h_3/(\lambda - I_3)$ $(i=1,2)$. One uses these relations in Eq. 16, which then gives $h_3/(\lambda - I_3)$ as a function of $2T$, of the form $k_3\sqrt{2T}$; or $\lambda = I_3 + h_3/k_3\sqrt{2T}$. Returning with this λ-value to recalculate ω_1^*/ω_3^* and ω_2^*/ω_3^*, we find that these ratios are not independent of $2T$, contrary to hypothesis, unless $k_1 = k_2 = 0$. In short, we find that Euler's problem leads to permanent axes that are truly inherent properties of the body only in the case of axial gyrostats, where $\underline{h}$ is aligned with a principal axis of inertia. The argument can be modified to consider ratios with respect to ω_1^* or ω_2^*, without changing this conclusion. In all such cases the permanent axes are principal axes of inertia. However, unlike the rigid body, not all principal axis are permanent axes. In fact, with the single exception of the "cylinder transverse subcase", the only principal axes that are permanent axes in Euler's original sense are those aligned with the $\underline{h}$ - vector.

While spin can occur about any osculation vector at a fixed direction in body axes, it can occur only at a specific angular velocity unless the osculation vector is a permanent axis.

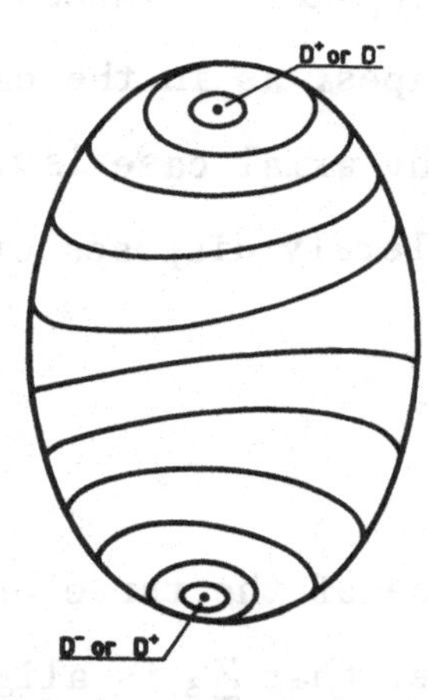

Fig. 3. Two Real Osculation Points

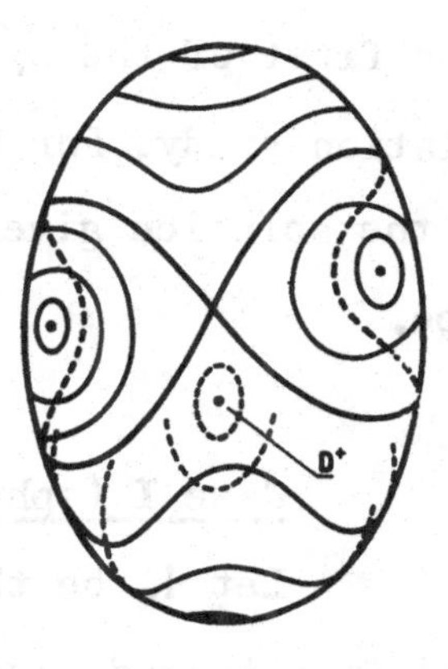

Fig. 4. Four Real Osculation Points

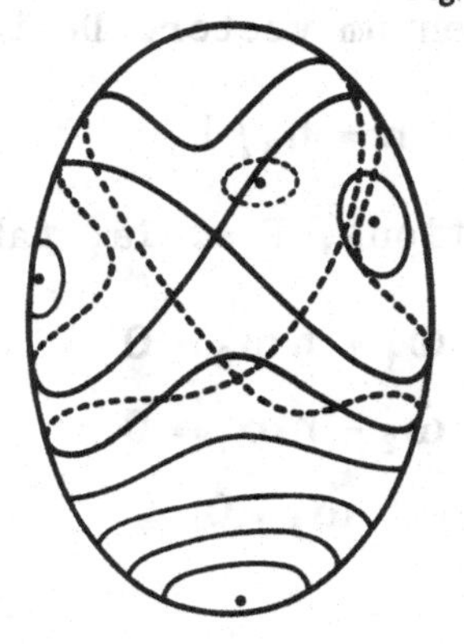

Fig. 5. Six Real Osculation Points

5. Axial Gyrostats, Canonical Cases

We now have seen the qualitative nature of the osculation vectors and the polhode families on the energy ellipsoid, for all body shapes. We now turn to the development of analytical solutions for the angular velocity components. In the present section, we confine ourselves to the special case of the axial gyrostat for which $h_1 = h_2 = 0$, $h_3 > 0$. It is convenient to dispose first of the special body shapes, as in the case of the osculation study. For the Sphere, the axial case is the only case, so the solution given for it completely disposes of that body shape.

<u>Case I (sphere)</u>

Let I_0 be the common value of the three principal moments of inertia and assume, as before, that $\underline{X}_3$ is aligned with the internal angular momentum vector. Define

$$n = h_3 / I_0 \tag{30}$$ (*)

The Euler dynamical equations, Eqs. 1a, take the simple form

$$\dot{\omega}_1 + n\omega_2 = 0 \tag{31a}$$

$$\dot{\omega}_2 - n\omega_1 = 0 \tag{31b}$$

$$\dot{\omega}_3 = 0 \tag{31c}$$

(*) Equation numbers are continued from Lecture 4.

From these,

$$\omega_1 = \omega_{10} \cos nt - \omega_{20} \sin nt \tag{32a}$$

$$\omega_2 = \omega_{20} \cos nt - \omega_{10} \sin nt \tag{32b}$$

$$\omega_3 = \omega_{30} \tag{32c}$$

<u>Case II A (cylinder, axial subcase)</u>

With $I_1 = I_2$ and $h_1 = h_2 = 0$, the Euler equations are still of the form of Eqs. 31 except that now

$$n = \left[h_3 + (I_3 - I_1)\omega_{30}\right] / I_2 \tag{33}$$

We could, of course, replace I_2 by I_1 in the formula, but we need the more general definition later. It follows, then, that the solutions for this subcase are still given by Eqs. 32.

Hereafter, we assume that the three principal moments of inertia are all different, so we are dealing with the Ellipsoid Case. For notational convenience, we replace I_1, I_2, and I_3 by A, B, and C respectively. In developing canonical forms of the solution we label axes such that B is the <u>intermediate</u> moment of inertia. This is somewhat analoguous to the ordering principle used to give canonical forms in the case of the rigid body, where we assumed the labeling was such that $C < B < A$, but now we admit either of two possible body shapes:

Case III AO(oblate axial ellipsoid) $A < B < C$ (34a)

Case III AP(prolate axial ellipsoid) $C < B < A$ (35b)

Previously we could aubsume both of these with one case, for the

$\underline{X}_1$ and $\underline{X}_3$ axes were interchangeable merely by relabeling. Now, however, the $\underline{X}_3$ label already is preempted by the direction of the internal angular velocity vector.

The osculation vectors for this case are summarized in Table 2. The derivation of these results from the cited equations is straightforward. All polhodes are swept out as D goes from its minimum to its maximum value. To develop canonical solution forms, we need a way to sweep them out unambiguously by a proper choice of initial conditions. All polhodes, in this axial case, cut the plane $\omega_{20}=0$ at some point, suggesting that canonical initial conditions should include $\omega_{20}=0$. If we put

(35) $$\omega_{10}=\sqrt{2T/A}\,\cos\Theta\,,\quad \omega_{30}=\sqrt{2T/C}\,\operatorname{sgn}(A-C)\sin\Theta$$

we are assured of satisfying the energy integral identically. (The use of the signum function in the definition of ω_{30} anticipates the path of greatest convenince later.) The initial canonical state and the corresponding polhode are uniquely characterized by θ if we regard $2T$ as a known constant throughout. The corresponding value of D is easily found from the momentum integral to be

(36) $$D = A\cos^2\Theta + C\left(\sin\Theta + \eta\,\operatorname{sgn}(A-c)/C\right)^2$$

where

(37) $$\eta = h_3\sqrt{C/2T}$$

Table 2 Osculation Vectors (Case III A (axial ellipsoid))

Ways to satisfy Eq. 15

Eq. 15 satisfied by	$\lambda = A, \quad \omega_2^* = 0,$ $\omega_3^* = h_3/(A-C)$	$\lambda = B, \quad \omega_1^* = 0,$ $\omega_3^* = h_3/(B-C)$	$\omega_1^* = \omega_2^* = 0$
Energy integral (Eq. 3b) gives	$\pm\omega_1^*$	$\pm\omega_2^*$	$\pm\omega_3^*$
Generic form of osculation vectors	$\left[\pm\omega_1^* \quad 0 \quad \omega_3^*\right]^T$	$\left[0 \quad \pm\omega_2 \quad \omega_3^*\right]^T$	$\left[0 \quad 0 \quad \pm\omega_3^*\right]^T$
Limitation: solution exists provided	$h_3 < \lvert A-C\rvert \sqrt{2T/C}$	$h_3 < \lvert B-C\rvert \sqrt{2T/C}$	no limitation
D-values from Eq. 3a	$D_A = A + \dfrac{Ah_3^2/2T}{(A-C)}$	$D_B = B + \dfrac{Bh_3^2/2T}{(B-C)}$	$D_{CU} = \left(\sqrt{C} + h_3/\sqrt{2T}\right)^2$ $D_{CL} = \left(\sqrt{C} - h_3/\sqrt{2T}\right)^2$

It follows that

(38a) $$D - D_A = (C - A)\left[\sin\Theta - \eta/|C-A|\right]^2$$

(38b) $$D - D_B = (A - B)\cos^2\Theta + (C - B)\left[\sin\Theta - \eta/|C-B|\right]^2$$

(Note that $\text{sgn}(A-C) = \text{sgn}(B-C)$.) An elementary investigation of the properties of $D(\Theta)$ given by Eq. 36 reveals that

(39a) $$D(\pm\pi/2) = \left(\sqrt{C} \pm \eta\,\text{sgn}(A-C)/\sqrt{C}\right)^2$$

and that if $\eta < |A-C|$, $D(\Theta)$ has an extremum D_{ex} at $\sin\Theta = \eta/|A-C|$, with

(39b) $$D_{ex} = A + A\eta^2/C(A-C)$$

It is a detail to show that the range of D-values that sweeps out all polhodes is defined by

(40) $$D(-\pi/2)\,\text{sgn}(A-C) \leq D\,\text{sgn}(A-C) \leq \begin{cases} D_{ex}\,\text{sgn}(A-C) & \eta \leq |A-C| \\ D(\pi/2)\,\text{sgn}(A-C) & \eta \geq |A-C| \end{cases}$$

Return to Eq. 3, multiplying the energy integral by B and subtracting it from the momentum integral. The result can be written

(41) $$\omega_1^2 = \frac{C(B-C)}{A(A-B)}\left[(\omega_3 + h_3/(C-B))^2 + R\right]$$

where

$$R = 2T(D - D_B)/C(B - C) =$$

$$= \frac{2T}{C}\left[\frac{A-B}{B-C}\cos^2\Theta - (\sin\Theta - \eta/|C-B|)^2\right] \tag{42}$$

The algebraic sign is of central importance. Define

$$\eta^+ = \sqrt{(A-C)(B-C)} \tag{43}$$

Then:

1. $R<0$ for $\eta > \eta^+$
2. there is a range of Θ such that $R \geq 0$ for $\eta \leq \eta^+$ provided $A \neq B$. Specifically, let

$$\sin\begin{Bmatrix}\Theta^-\\ \Theta^+\end{Bmatrix} = \frac{\eta}{|A-C|} \mp \sqrt{\frac{A-B}{A-C}\left[1-(\eta/\eta^+)^2\right]} \tag{44}$$

Then $R \geq 0$ for $\Theta^- \leq \Theta \leq \Theta^+$ while $R<0$ for other Θ -values.

It is the change of algebraic sign for $R(\Theta)$ that distinguishes solutions of the two basic types for the axial gyrostat. We call them

endcap solution, $R < 0$

sidecap solution, $R > 0$

The separation $R=0$ corresponds to $D = D_B$, by Eq. 45. The polhode $D = D_B$ is a separatrix which divides the polhodes into two classes.

Again return to Eqs. 3, this time multiplying the energy integral by A and subtracting it from the momentum integral to get

$$-\omega_2^2 = \frac{C(A-C)}{B(A-B)}\left[(\omega_3 + h_3/(C-A))^2 - P^2\right] \tag{45}$$

where

$$P = \sqrt{2T|D_A - D|/C|A-C|} = \sqrt{2T/C}\,|\sin\Theta - \eta/|C-A|| \tag{46}$$

The right side of Eq. 45 can be decomposed into a product of linear factors with real zeros. We define

$$\varrho_1 = h_3/(A-C) + P \tag{47a}$$

$$\varrho_4 = h_3/(A-C) - P \tag{47b}$$

noting in passing that $\varrho_1 > \varrho_4$ and that either ϱ_1 or ϱ_4 equals ω_{30}, and write

$$\omega_2 = \pm\sqrt{\frac{C(A-C)}{B(A-B)}}\sqrt{(\varrho_1 - \omega_3)(\omega_3 - \varrho_4)} \tag{48}$$

On the other hand, the right side of Eq. 41 cannot be decomposed into linear factors with real zeros except for Θ -values such that $R=0$. For the moment, therefore, we simply write

$$\omega_1 = \pm \sqrt{\frac{C(B-C)}{A(A-B)}} \; \sqrt{(\omega_3 + h_3/(C-B))^2 + R} \tag{49}$$

Having now expressed both ω_1 and ω_2 as functions of ω_3, we return to the third row of Eq. 1a for the differential equation satisfied by ω_3, namely

$$\dot{\omega}_3 = (A-B)\,\omega_1\omega_2/C \tag{50}$$

Thus we have

$$\dot{\omega}_3 = \varepsilon \sqrt{\frac{(A-C)(B-C)}{AB}} \; \sqrt{(\varrho_1-\omega_3)\,(\omega_3-\varrho_4)\left[(\omega_3+h_3/(C-B))^2+R\right]} \tag{51}$$

where $\varepsilon = \pm 1$. To resolve the algebraic sign ambiguity, consider $\dot{\omega}_3$ at time Δt after $t=0$. The sign of ω_1 must be that of ω_{10} if Δt is sufficiently small, while ω_2 is approximated by the initial value of $\dot{\omega}_2 \Delta t$, the latter gotten form the second row of Eq. 1a. By Eq. 50, $\operatorname{sgn} \dot{\omega}_3 = \operatorname{sgn}\left[(A-B)\omega_1\omega_2\right]$. Hence

$$\begin{aligned} \varepsilon &= \operatorname{sgn}\left\{(A-B)\omega_{10}\left[(C-A)\,\omega_{10}\,\omega_{30} + h_3\,\omega_{10}\right]\right\} = \\ &= \operatorname{sgn}(A-C)\,\operatorname{sgn}\left(-\,|C-A|\sin\Theta + \eta\right) \end{aligned} \tag{52}$$

From this point we take up the endcap and sidecap solutions separately.

Endcap Solutions for the Axial Gyrostat

The four real roots of the quartic under the radical in Eq. 51 can be summarized as follows: ϱ_1 and ϱ_4 are given by Eqs. 47, with $\varrho_1 > \varrho_4$; the remaining two are

$$\left.\begin{matrix}\varrho_2\\ \varrho_3\end{matrix}\right\} = \sqrt{\frac{2T}{C}}\left(\eta/(B-C) \pm \sqrt{-R}\right) \tag{53}$$

Where the upper sign denotes ϱ_2 , the lower ϱ_3 , so $\varrho_2 > \varrho_3$. In these terms Eq. 51 can be rewritten in differential form as

$$\frac{d\omega_3}{\sqrt{-(\omega_3-\varrho_1)(\omega_3-\varrho_2)(\omega_3-\varrho_3)(\omega_3-\varrho_4)}} = \varepsilon\sqrt{\frac{(A-C)(B-C)}{AB}}\,dt \tag{54}$$

and the solution gotten by quadrature. To cast the solution in a simple canonical form we must consider the ordering of the roots ϱ_i in some detail. When they are ordered in increasing size, we denote them symbolically by

$$\delta < \gamma < \beta < \alpha \tag{55}$$

according to the convention used in Pierce's tables of 1929, pp. 70-71 . For brevity, denote the ordering $\varrho_4 < \varrho_3 < \varrho_2 < \varrho_1$ simply by 4321 , and analogously for other orderings. Figure 6 shows the way the root orderings changes with increasing η for various initial conditions and for the two possinilities $A<C$ and $A>C$. The depiction is purely schematic, and no significance should be attached to the exact shape of the region's boundaries. There is no problem in finding the conditions that determining the boundaries, if this is desired. Specifically, $\Theta = \Theta^-$ and $\Theta = \Theta^+$ from Eq.44fix the boundaries of the $R > 0$ region. The boundaries marked "BDY" in the figure occurs when $D = D_{CL}(C > A)$ or $D = D_{CU}$ $(A > C)$, and has the equation

$$\sin \Theta_0 = -1 + 2\eta / |C-A| \qquad (56)$$

The heavy dashes lines in the η-range $[\eta^+, |A-C|]$ have the equations

$$\sin \Theta_1 = \eta / |C-A| \qquad (57)$$

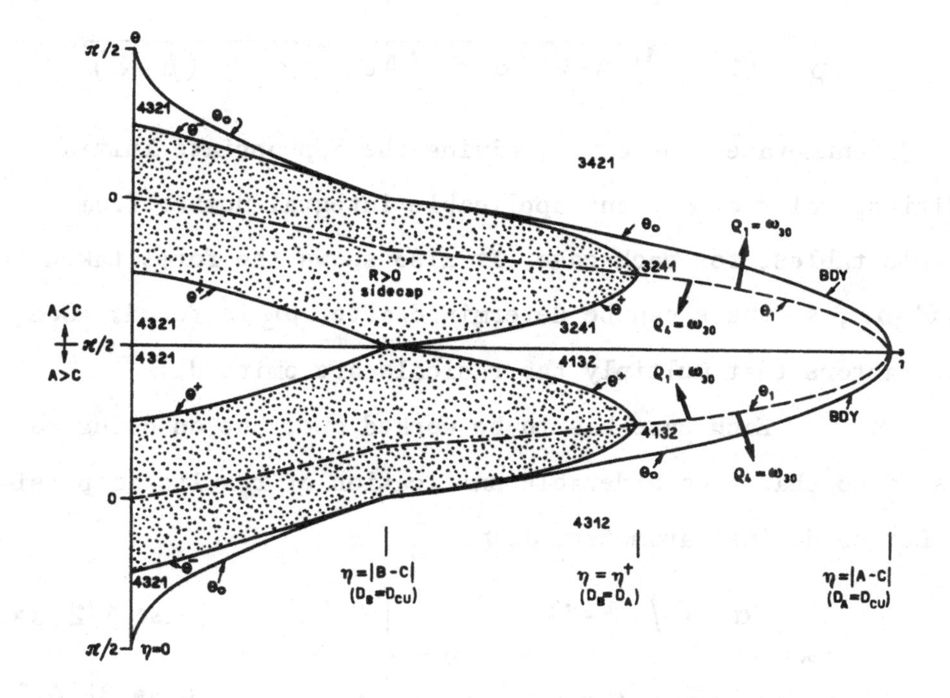

Fig. 6. Schematic Regions for Root Ordering

They, toghether with the $R>0$ region, separate the initial condition regions. Bold arrows in the figure designate the sides of these boundaries corresponding to $\varrho_1 = \omega_{30}$ and $\varrho_4 = \omega_{30}$.

We now consider the various cases that can arise. First rewrite the solution of Eq. 54 as

$$\int_{\omega_{30}}^{\omega_3} \frac{dx}{\sqrt{-(x-\alpha)(x-\beta)(x-\gamma)(x-\delta)}} = \varepsilon \sqrt{\frac{(A-C)(B-C)}{AB}}\; t \tag{58}$$

and define an elliptic function modulus k and a frequency p by

$$k = \sqrt{(\alpha-\beta)(\gamma-\delta)/(\alpha-\gamma)(\beta-\delta)} \tag{59}$$

$$p = (1/2)\sqrt{(A-C)(B-C)/AB}\;\sqrt{(\alpha-\gamma)(\beta-\delta)} \tag{60}$$

Table 3 enumerates the cases, giving the appropriate initial condition, value of ε, and applicable integral number from Pierce's tables, for each case. It also shows the forms taken by ω_1 and ω_2, so these can be determined after ω_3 is found. (Constant factors that multiply the radicals are omitted.)

Each case has to be worked out, but on doing so it is found that a considerable unification of results is possible. Let us define parameters u, v by

$$u = \begin{cases} (\alpha-\beta)/(\beta-\delta) \\ (\gamma-\delta)/(\alpha-\gamma) \end{cases} \qquad v = \begin{cases} \delta/\alpha & \text{Case 1, 2, 3a} \\ \alpha/\delta & \text{Case 3b, 4, 5} \end{cases} \tag{61}$$

Table 3 Endcap Solution Cases

Case	Root order	ω_{30}	ε	Applicable PEIRCE #	$\omega_1 \propto$	$\omega_2 \propto$
1a	3241	α	−1	551	$\sqrt{(\omega_3-\gamma)(\omega_3-\delta)}$	$\sqrt{(\alpha-\omega_3)(\omega_3-\beta)}$
1b		β	+1	552		
2	3421	α	−1	551	$\sqrt{(\omega_3-\beta)(\omega_3-\beta)}$	$\sqrt{(\alpha-\omega_3)(\omega_3-\gamma)}$
3a	4321	α	−1	551	$\sqrt{(\omega_3-\beta)(\omega_3-\gamma)}$	$\sqrt{(\alpha-\omega_3)(\omega_3-\delta)}$
3b		δ	+1	556		
4	4312	δ	+1	556	$\sqrt{(\alpha-\omega_3)(\gamma-\omega_3)}$	$\sqrt{(\beta-\omega_3)(\omega_3-\delta)}$
5a	4132	δ	+1	556	$\sqrt{(\alpha-\omega_3)(\beta-\omega_3)}$	$\sqrt{(\gamma-\omega_3)(\omega_3-\delta)}$
5b		γ	−1	555		

and dimensionless time τ by

$$\tau = pt + \begin{cases} K & \text{Cases 1b,5b} \\ 0 & \text{otherwise} \end{cases} \tag{62}$$

Then in all cases, ω_3 becomes

$$\omega_3 = \omega_{30}\,(1 + \upsilon u\ \mathrm{sn}^2\tau)/(1 + u\ \mathrm{sn}^2\tau) \tag{63}$$

(The modulus k is implicit in the elliptic functions).

Three different forms occur for ω_1 and ω_2. They are tabulated in Table 4. The coefficients in these results can be gotten in terms of the roots ϱ_i ; indeed, this is the way they must naturally appear when Eq. 63 is substituted into Eqs. 48 and 49. However, it is both simpler and less likely to lead to numerical error to replace the coefficients by their equivalents in terms of initial conditions. It is obvious that both ω_3 given by Eq. 63 and ω_1 given by Table 4 have the proper initial values. The coefficient for ω_2 in Table 4 is selected so the derivative $\dot{\omega}_2$ matches the initial value of the derivative gotten from Eq. 1a. In Table 4, the u-value is the one appropriate to the case in question. The n-value is from Eq. 33.

Note that as $A \to B$, both $k \to 0$ and $u \to 0$. Also $\varrho_2 \to \varrho_1$ and $\varrho_2 \to \varrho_1$, so only cases 2 and 3 or Table 4 are possible. In both of these, the solutions approach

$$\omega_1 = \omega_{10}\cos pt\,,\quad \omega_2 = (\omega_{10} n/p)\sin pt\,,\ \omega_3 = \omega_{30} \tag{64}$$

Table 4 Solutions for ω_1, ω_2

Solution type	Cases	ω_1/ω_{10}	$\omega_2/(\omega_{10}\, n/p)$
1	1,5	$\dfrac{\mathrm{dn}\,\tau}{1+u\,\mathrm{sn}^2\tau}$	$\dfrac{\mathrm{sn}\,\tau\ \mathrm{cn}\,\tau}{1+u\,\mathrm{sn}^2\tau}$
2	2,4	$\dfrac{\mathrm{dn}}{1+u\,\mathrm{sn}^2\tau}$	$\dfrac{\mathrm{sn}\,\tau\ \mathrm{dn}\,\tau}{1+u\,\mathrm{sn}^2\tau}$
3	3	$\dfrac{\mathrm{cn}\,\tau\ \mathrm{dn}\,\tau}{1+u\,\mathrm{sn}^2\tau}$	$\dfrac{\mathrm{sn}\,\tau}{1+u\,\mathrm{sn}^2\tau}$

Furthermore, it can be shown that p approaches the value n given by Eq. 33, whence the solution reduces to that given previously for the "cylinder, axial subcase".

An algorithmic solution of the endcap case can be summarized as follows:

1. input data are A, B, C, h_3, Θ and t-range
2. calculate ω_{10}, ω_{30} (Eq. 35), ε (Eq. 52), n (Eq. 33), $2T$ (Eq. 3b), η (Eq. 37)
3. calculate ϱ_1, ϱ_4 (Eq. 47)
4. calculate R (Eq. 42). If $R \leq 0$ continue with endcap solution
5. calculate ϱ_2, ϱ_3 (Eq. 53)
6. order ϱ_i in increasing order

7. determine the applicable case of Table 3.

A simple logic for this step, if it is to be mechanized, is based on an array IX[1:4], the first element of which is the index of the smallest ϱ-value, the second of which is the index of the next larger ϱ-value, etc. This array is easily constructed when the roots are ordered in step 6. Then if CASE is the case in Table 4,

```
CASE : IF (IX[1] + IX[2]  EQL 5) THEN 4*IX[4]-3
                 ELSE 8-IX[2]-IX[3];
```

8. determine the solution type in Table 4. Note that TYPE:
= 3 - ABS (CASE -3);

9. calculate the values of u and v (Eqs. 61). If RHO[1:4] is the array of ϱ-values after ordering, i.e. RHO[1]=δ, RHO[2]=γ etc. in the notation used in Eq. 55 ff, a simple logic is:

```
I := IF (CASE + EPSILON LEQ 2) THEN 2 ELSE 3;
V:= RHO [10 -3*I]; TEMP: = RHO [3*I -5];
U:= (TEMP-RHO [I])/(RHO [I] - V);
V:= V / TEMP;
```

10. calculate k (Eq. 59) and p (Eq. 60)

11. by a suitable procedure, use k to calculate K, the complete elliptic integral of the first kind

12. use p and the input t-range, construct the corresponding τ-range, adding K to the results in Cases 1b and 5b

(Eq. 62), according to the logic

IF ABS (CASE -3- EPSILON) GEQ 3 THEN TAU: = TAU+ K;

(following THEN would be a FOR-statement if this addition were to be done to all of an array of τ-values.)

13. by a suitable procedure, calculate $\mathrm{sn}(\tau,k)$, $\mathrm{cn}(\tau,k)$ and $\mathrm{dn}(\tau,k)$
14. for each τ of the set desired, compute ω_3 (Eq. 63) and ω_1, ω_2 according to the row of Table 4 specified by variable TYPE

If complete polhodes are to be constructed, steps 1-11 need be done but once, while steps 12-14 are done for each calculated point of the polhode.

Sidecap Solutions for the Axial Gyrostat

To develop canonical sidecap solutions, we follow the method described by Whittaker and Watson (1927; 22.71, 22.72). The basic problem is to deal with Eq. 51 for the case where two roots of the quartic are complex. Consider the two quadratic components of the quartic,

$$S_1 = (x + h_3/(C-B))^2 + R, \quad S_2 = P^2 - (x + h_3/(C-A))^2 \qquad (65)$$

(x temporarily represents ω_3) with $R > 0$. The quadratic form $S_1 - \nu S_2$ is a perfect square if ν is chosen to make the discriminant of the form equal to zero, i.e. if ν satisfies

$$P^2\nu^2 + \left[P^2 - R - (A-B)^2 h_3^2/(C-A)^2(C-B)^2\right]\nu - R = 0 \tag{66}$$

Denote the two roots of this equation by ν_1 and ν_2 with $\nu_1 > \nu_2$ It is evident that when $R>0$ these roots are real, and that they have the properties $\nu_1 > 0 > \nu_2$, $1+\nu_1 > 1+\nu_2 > 0$. We find that

$$S_1 - \nu_i S_2 = (1+\nu_i)(x+\alpha_i)^2 \qquad (i = 1,2) \tag{67}$$

where

$$\alpha_i = h_3\left[1/(C-B) + \nu_i/(C-A)\right]/(1+\nu_i) \tag{68}$$

(Here α is not related to the root denoted α in the last subsection). Finally, solving Eqs. 67 for S_1 and S_2, the two quadratic forms become

$$S_1 = \left[\nu_1(1+\nu_2)(x+\alpha_2)^2 - \nu_2(1+\nu_1)(x+\alpha_1)^2\right]/(\nu_1-\nu_2) \tag{69a}$$

$$S_2 = \left[(1+\nu_2)(x+\alpha_2)^2 - (1+\nu_1)(x+\alpha_1)^2\right]/(\nu_1-\nu_2) \tag{69b}$$

Now replace x by ω_3 and change variables by the transformation

$$y = \sqrt{(1+\nu_1)(1+\nu_2)}\,(\omega_3+\alpha_1)/(\omega_3+\alpha_2) \tag{70}$$

Equation 51 becomes, in terms of this new variable,

$$\dot{y} = \frac{(\alpha_2-\alpha_1)\sqrt{(1+\nu_1)(1+\nu_2)}}{(\nu_1-\nu_2)}\,\varepsilon\sqrt{\frac{(A-C)(B-C)}{AB}}\times \tag{71}$$

$$\times \sqrt{(1-y^2)(\nu_1 + |\nu_2| y^2)} \tag{71}$$

Now define

$$p = \varepsilon(\alpha_1 \mp \alpha_2)\sqrt{\frac{(1+\nu_1)(1+\nu_2)}{(\nu_1-\nu_2)} \frac{(A-C)(B-C)}{AB}} \tag{72}$$

$$k = \sqrt{|\nu_2| / (\nu_1 - \nu_2)} \leq 1 \tag{73}$$

and recognize that one of the values $y = \pm 1$ is an initial condition of the problem, for these are the roots of S_2 and hence (see Eq. 51) the places where ω_2 becomes zero. More specifically, if $(\Theta - \Theta_1)(A-C) > 0$ the initial condition is $\omega_3 = \varrho_1$ or $y = -1$, whereas if the inequality is opposite the initial condition is $\omega_3 = \varrho_4$ or $y = +1$. We therefore rewrite Eq. 71 as

$$\int_{\pm 1}^{y} \frac{dy}{\sqrt{(1-y^2)(k'^2 + k^2 y^2)}} = -pt \tag{74}$$

or

$$\int_{y}^{1} \frac{dy}{\sqrt{(1-y^2)(k'^2+k^2y^2)}} = pt + \int_{\pm 1}^{1} \frac{dy}{\sqrt{(1-y^2)(k'^2+k^2y^2)}} = pt + \begin{Bmatrix} 0 \\ 2K \end{Bmatrix} \tag{75}$$

where K is the complete elliptic integral of the first kind. Finally, using Peirce #529,

$$y = \begin{cases} \text{cn}\,(pt, k) \\ \text{cn}\,(pt + 2K, k) \end{cases} = \pm\,\text{cn}\,pt. \tag{76}$$

(As before, the modulus is omitted for brevity.)

Returning from y to ω_3 by way of Eq. 70, and recalling the definition of ε by Eq. 52,

$$\omega_3 = \frac{\varepsilon\,\alpha_2\sqrt{1+\nu_2}\ \text{cn}\,pt - \alpha_1\sqrt{1+\nu_1}}{\sqrt{1+\nu_1} - \varepsilon\sqrt{1+\nu_2}\ \text{cn}\,pt}\,. \tag{77}$$

It is somewhat more convenient to express ω_3 in terms of $1-\text{cn}\,pt$ so a certain combination of α_1, α_2, ν_1, and ν_2 can be recognized immediately as the initial value ω_{30} and possible numerical error can be obviated. If this be done, defining

$$u = \left(1 - \varepsilon\sqrt{(1+\nu_1)\,(1+\nu_2)}\right)^{-1} \tag{78}$$

we get

$$\omega_3 = \frac{\omega_{30} + u\alpha_2\,(1-\text{cn}\,pt)}{1 - u\,(1-\text{cn}\,pt)} \tag{79}$$

With these solutions for ω_3 and y, we can return to the forms S_1 and S_2 and show that they are proportional to certain time functions:

$$\omega_1 \propto S_1 \propto \text{dn}\,pt/[1-u\,(1-\text{cn}\,pt)] \tag{80a}$$

$$\omega_2 \propto S_2 \propto \text{sn } pt / [1 - u(1 - \text{cn } pt)] \tag{80b}$$

As in the case of the endcap solution, it is just a matter of choosing the proportionality constants so the initial values of ω_1 and $\dot{\omega}_2$ are correct. The final results are

$$\omega_1 = \omega_{10} \text{ dn } pt \ / \ [1 - u(1 - \text{cn } pt)] \tag{81}$$

$$\omega_2 = (\omega_{10} n/p) \text{ sn } pt / [1 - u(1 - \text{cn } pt)] \tag{82}$$

An algorithmic solution of the sidecap case can be summarized as follows, branching from the endcap algorithm above after step 4:

5. calculate ν_1, ν_2 (Eq. 66), α_1, α_2 (Eq. 68)
6. calculate p (Eq. 72), k (Eq. 73), u (Eq. 78)
7. by a suitable procedure, calculate sn pt, cn pt, dnpt for the t-values of concern
8. for each t-value, compute ω_3 (Eq. 79), ω_1 (Eq. 81), ω_2 (Eq. 82)

6. The General Free Gyrostat : Recent Investigations

In two previous lectures the gyroscopic system called gyrostat has been presented and two kinds of problem connected with the case of torque-free motion have been discussed. In the following we will consider the torque-free motion of the unsymmetrical gyrostat with a rotor which is not aligned with a principal axis nor does it lie in the plane of two principal axes. Since the polhodes, i.e. the intersection lines of the energy and the angular momentum ellipsoids allow a simple visualization of the motion we will start with a discussion of some interesting problems concerning the polhodes. In particular, we return to the so-called osculation problem the solution of which provides the possible states of permanent rotation. The condition for such a state of motion has already been developed. It results in the equations

$$\sum_{\alpha=1}^{3} \frac{h_\alpha^2 I_\alpha}{(I_\alpha - \lambda)} = 2T \tag{1}$$

$$\omega_\alpha = \frac{-h_\alpha}{I_\alpha - \lambda}, \quad \alpha = 1,2,3 \tag{2}$$

Any real solution λ of (1) yields three components of a permanent angular velocity ω according to (2). On the energy ellipsoid

$$\sum_{\alpha=1}^{3} I_\alpha \omega_\alpha^2 = 2T \tag{3}$$

the permanent angular velocities appear as single points. These points determine how the polhodes of all other possible motions for which $2T$ has the same value are arranged on the surface of the ellipsoid. If a permanent rotation is stable then the point is surrounded by closed polhodes. If it is unstable it is the crossing point of a particular polhode which is called separatrix because it separates the regions of influence of stable permanent rotations (s. Fig. 1). These few remarks show that for an understanding of the dynamics of the gyrostat it is important to know how many axes of permanent rotation exist. In other words: How many real roots λ does equ. (1) have? This problem which will be considered in detail now is rather complicated because the number of real roots is a function of many parameters, namely of $I_{1,2,3}$, $h_{1,2,3}$ and $2T$.

It is convenient to introduce the absolute value h of the relative rotor angular momentum and the direction cosines $u_{1,2,3}$ of the rotor axis in the principal axes frame of reference, i.e. $h_\alpha = h u_\alpha$ $(\alpha = 1,2,3)$. Then, (1) can be rewritten in the form

$$F(\lambda) = \sum_{\alpha=1}^{3} \frac{u_\alpha^2 I_\alpha}{(I_\alpha - \lambda)^2} = \frac{2T}{h^2} - \frac{1}{I_0} \tag{4}$$

where now all terms on the left side are design parameters and the terms on the right side are state of motion parameters. Thus, the number of real roots has to be expressed as a function of the moments of inertia $I_{1,2,3}$, of the rotor axis direction $u_{1,2,3}$ and of the newly defined parameter I_0 (dimension of a moment of inertia). The direction cosines obey the condition

$$\sum_{\alpha=1}^{3} u_\alpha^2 = 1 \tag{5}$$

Before entering any mathematical development we can make a general statement based on the knowledge of the behaviour of the rigid body without a rotor. If an arbitrary gyrostat with three different moments of inertia and with some arbitrary value $2T$ is given then we can predict that with h approaching zero (very slow relative rotor motion) the gyrostat behaves more and more like a rigid body. Consequently, in the limit, it has axes of permanent rotation coinciding with its principal axes of inertia. They intersect the energy ellipsoid in six points. If, on the other hand, the relative motion becomes faster and faster then the dynamical behaviour is dominated by the rotor. The slow carrier body has no influence any more. Consequently, in the limit, the gyrostat becomes a symmetrical rigid body. It then is able to carry out permanent rotations about its axis of symmetry (the rotor axis) which intersects the energy ellipsoid in two points. Thus, we can predict that no matter what $I_{1,2,3}$ and $u_{1,2,3}$ are (Eq. 4) will have six real roots for $I_0 \to 0$ and two real roots for $I_0 \to \infty$. This statement follows also immediately from (4) without reference to physical arguments. In the assumed case where $I_{1,2,3}$ are all different and $u_{1,2,3}$ are all non-zero, Eq. 4 has no root $\lambda = I_\alpha$ $(\alpha = 1,2,3)$ and, consequently, it can be given the form of a sixth-order polynomial. We thus have either six or four or two or zero real roots. The function $F(\lambda)$ defined by (4) is positive with $F(\lambda) \to 0$ for $\lambda \to \pm\infty$. This means that for $I_0 > 0$ the case of zero real roots does not exists.

$F(\lambda)$ has second-order poles at $\lambda = I_1$, $\lambda = I_2$ and $\lambda = I_3$. It has no inflection points since $\frac{\partial^2 F}{\partial \lambda^2} > 0$. This means that $F(\lambda)$ has exactly two minima, one in the range $I_3 < \lambda < I_2$ and one in the range $I_2 < \lambda < I_1$ (without loss of generality it is assumed that $I_3 < I_2 < I_1$). It is, therefore, represented by a curve shown schematically in Fig. 2. From these properties it follows directly that, indeed, the equation $F(\lambda) = 1/I_0$ has six real roots for $I_0 \to 0$ and two real roots for $I_0 \to \infty$. Fig. 2 also shows that, in general, in a certain range of values of I_0 (all other parameters kept fixed) the equation has four real roots. This range of I_0 is easily found. The transition from six to four or from four to two real roots of (4) happens when (4) has double roots, i.e. if $\frac{\partial F}{\partial \lambda} = 0$ or

$$\sum_{\alpha=1}^{3} \frac{u_\alpha^2 I_\alpha}{(I_\alpha - \lambda)^3} = 0 . \tag{6}$$

This, again, is a sixth-order polynomial. From the properties of $F(\lambda)$ we know that it has exactly two real roots, one in the range $I_3 < \lambda < I_2$ and one in the range $I_2 < \lambda < I_1$. Substitution of these roots into (4) yields two values for I_0 which determine the range of I_0 in which (4) has four real solutions.

Next we turn to the more difficult problem of the relationship between the number of real roots of (4) and the rotor axis direction $u_{1,2,3}$. Now I_0 is supposed to be given and the function $F(\lambda)$ is changing with its parameters. Again, the transition from six to four or from four to two real solutions of (4) occurs when (6) is fulfilled. Equations (4), (5) and (6) read

$$(7)\qquad \begin{cases} \dfrac{I_1}{(I_1-\lambda)^2}\, u_1^2 + \dfrac{I_2}{(I_2-\lambda)^2}\, u_2^2 + \dfrac{I_3}{(I_3-\lambda)^2}\, u_3^2 = \dfrac{1}{I_0} \\ \dfrac{I_1}{(I_1-\lambda)^3}\, u_1^2 + \dfrac{I_2}{(I_2-\lambda)^3}\, u_2^2 + \dfrac{I_3}{(I_3-\lambda)^3}\, u_3^2 = 0 \\ u_1^2 + u_2^2 + u_3^2 = 1 \end{cases}$$

This set constitutes three linear equations for the squares u_1^2, u_2^2, u_3^2 in terms of parameter λ. The physical interpretation of the solution $u_{1,2,3}^2(\lambda)$ is the following. Suppose that I_0, I_1, I_2 and I_3 are given parameters and that only the rotor direction in the carrier body is changing. Then we can assume, in general, that for some rotor directions the gyrostat will have six axes of permanent rotation, for others four and for still others only two. Thus, the unit vector along the rotor axis which has components $u_{1,2,3}$ will produce on the surface of a body-fixed unit sphere closed regions belonging to six, four or two real roots of (4), respectively. At the boundaries of these regions (7) is fulfilled. Consequently, the solution of (7) is a parameter representation of these boundary lines. It can be easily calculated. A convenient form is obtained with the help of the identities

$$(8)\qquad I_1^2\,(I_1-I_2) + I_2^2\,(I_3-I_1) - I_3^2\,(I_1-I_2) =$$

$$= I_2 I_3 (I_2 - I_3) + I_3 I_1 (I_3 - I_1) + I_1 I_2 (I_1 - I_2) = -(I_1 - I_2)(I_2 - I_3)(I_3 - I_1) \tag{8}$$

The result is

$$u_\alpha^2(\lambda) = \frac{I_\beta I_\gamma (I_\beta - I_\gamma)(\lambda - I_\alpha)^3 \left[\lambda^3/(I_\beta I_\gamma) - 3\lambda + I_\beta + I_\gamma - I_0\right]}{I_0 (I_1 - I_2)(I_2 - I_3)(I_3 - I_1)(\lambda^3 - I_1 I_2 I_3)} \tag{9}$$

$$\alpha = 1,2,3 \qquad \alpha, \beta, \gamma \text{ cyclic}$$

Boundary lines on the unit sphere actually exist only if there are values of λ for which all three squares $u_{1,2,3}^2$ are non-negative. Our initial arguments have shown that independent of $I_{1,2,3}$, for sufficiently small values I_0 only six and for sufficiently large values I_0 only two real solutions of (4) can exist no matter what the rotor direction is. This means that then no boundary lines occur. The entire surface of the unit sphere belongs to six or two solutions, respectively. For values I_0, however, that are neither excessively small nor excessively large large boundary lines can be expected. This raises two questions: First, how do the boundary lines look if there are any, and secondly, what are the minimum and maximum possible values of I_0 which allow boundary lines?

We start with the first problem. The structure of (9) in connection with the properties of the function $F(\lambda)$ allows the following statements. Values of λ for which all three squares $u_{1,2,3}^2$ are non-negative must lie in the ranges $I_3 < \lambda < I_2$

and/or $I_2 < \lambda < I_1$. In general, therefore, there will be one or two intervals for λ leading to $u^2_{1,2,3} \geq 0$. Suppose a finite λ--interval with these properties exists. Then, at both endpoints of this interval at least one of the squares $u^2_{1,2,3}$ is zero. This means that the corresponding point of the boundary line lies on one of the great circles represented by the equations $u_1 = 0$, $u_2 = 0$ and $u_3 = 0$, respectively. Between these two points the boundary line is continuous because the functions (9) are continuous. Because of the square on the left side of (9) the boundary lines are symmetric with respect to all three great circles on the unit sphere. Thus, any boundary line is a closed curve and one octant created by the great circles on the sphere describes the entire sphere. The actual calculation of boundary lines starts with the calculation of the roots of the three third-order polynomials of λ appearing in each of the functions (9). These roots determine the λ-intervals for which non-negative squares $u^2_{1,2,3}$ exist. Since these λ-intervals will be needed later we shall determine them next.

The factor $I_\beta I_\gamma / [I_0(I_1-I_2)(I_2-I_3)(I_3-I_1)]$ in (9) is negative for all three functions $u^2_\alpha(\lambda)$. The factor $(I_\beta - I_\gamma)$ is positive for $\alpha = 1$ and $\alpha = 3$ and it is negative for $\alpha = 2$. Writing (9) in the form

$$u^2_\alpha(\lambda) = \frac{I_\beta I_\gamma (I_\beta - I_\gamma)(\lambda - I)^3}{I_0(I_1-I_2)(I_2-I_3)(I_3-I_1)(\lambda^3 - I_1 I_2 I_3)} \cdot f_\alpha(\lambda), \tag{10}$$

$$\alpha = 1,2,3, \quad \alpha,\beta,\gamma \text{ cyclic} \tag{10}$$

with

$$f_\alpha(\lambda) = \lambda^3/(I_\beta I_\gamma) - 3\lambda + I_\beta + I_\gamma - I_0 \quad \alpha = 1,2,3 \tag{11}$$

the diagrams 1a,b can be drawn which show as functions of λ the algebraic signs of the factors in front of $f_\alpha(\lambda)$ in (10) for $\alpha = 1,2,3$. They are to be read as follows. The notation $\overset{3+}{\longleftrightarrow}$ means that the above mentioned factor $\alpha = 3$ is positive in the λ-range indicated by the arrows, i.e. in the case of the upper diagram 1a in the range $I_3 < \lambda < (I_1 I_2 I_3)^{1/3}$. Likewise $\overset{2-}{\longleftrightarrow}$ means that the above factor for $\alpha = 2$ is negative in the range indicated by these arrows etc. Two different diagrams are necessary because the cubic mean $(I_1 I_2 I_3)^{1/3}$ can be, both, smaller and larger than I_2 (the possible case where both quantities are equal needs no separate diagram. Notice that then the interval $\overset{2-}{\longleftrightarrow}$ vanishes). These diagrams have to be combined with similar ones in which the algebraic signs of the factors $f_\alpha(\lambda)$ are indicated. This superposition, then, furnishes the λ-intervals for non-negative squares $u^2_{1,2,3}$. Unfortunately, the roots of the polynomials (11) cannot be expressed analytically in a useful way. Therefore, at this point, in general, numerical calculations have to start. Before showing results of such calculations we shall now turn to the second problem defined earlier, i.e. to the determination of the minimum and maximum values I_0 for which boundary lines exist. Using the arguments of the previous discussion this can be formulated also as follows. What are the extreme values $I_{0\,min}$ and $I_{0\,max}$ for which a region on the unit sphere defined by a closed boundary line shrinks to a single point? Or still differnetly formulated: What are the extreme values $I_{0\,min}$ and $I_{0\,max}$ such that there exists only one single value λ (instead of one or two finite intervals) for which all three squares $u^2_{1,2,3}$ are non-negative? From this last formulation follows directly that in the critical cases of $I_{0\,min}$ and $I_{0\,max}$ the boundary lines are single points on the great circles on the sphere. This means that the rotor axis is either in the plane of two principal axes of inertia or even along a principal axis of inertia. Thus, our initial assumption of all three h-components being non-zero is violated and with it the basis of equations (9). We can, however, consider these critical solutions as limiting cases which are approached in a continuous manner by a sequence of solutions for which (9) holds. This was expressed above already when it was said that closed boundaries shrink to a single point.

We can now return to the diagrams 1a,b and to the discussion of the polynomials $f_\alpha(\lambda)$ of (11). Our present problem can then be formulated as follows. What are the extreme values $I_{0\,min}$ and $I_{0\,max}$ such that the algebraic sign distribution of the polynomials (11) superimposed on diagrams provides a single point λ for which all three squares $u^2_{1,2,3}$ are non-negative? This restricted problem can be solved without numerical calculations. We notice that the function $f_\alpha(\lambda)$ has extreme values at $\lambda = \pm\sqrt{I_\beta I_\gamma}$ (α, β, γ cyclic). At $\lambda = +\sqrt{I_\beta I_\gamma}$ it has a minimum of the magnitude

$$f_{\alpha\,min} = \left(\sqrt{I_\beta} - \sqrt{I_\gamma}\right)^2 - I_0 \tag{12}$$

It is easy to show that of the three minima $f_{2\,min}$ is the largest one. This follows directly form the inequalities $I_3 < I_2 < I_1$ which also mean that $\sqrt{I_3} < \sqrt{I_2} < \sqrt{I_1}$. The relative size of $f_{1\,min}$ and $f_{3\,min}$, on the other hand, is more complicated. The difference of the two is

$$\begin{aligned} f_{1\,min} - f_{3\,min} &= \left(\sqrt{I_2} - \sqrt{I_3}\right)^2 - \left(\sqrt{I_1} - \sqrt{I_3}\right) = \\ &= 2\sqrt{I_2}\left(\sqrt{I_1} - \sqrt{I_3}\right) - \left(I_1 - I_3\right) = \\ &= \left(\sqrt{I_1} - \sqrt{I_3}\right)\left[2\sqrt{I_2} - \left(\sqrt{I_1} + \sqrt{I_3}\right)\right] \end{aligned}$$

Consequently

$$f_{1\,min} - f_{3\,min} \gtrless 0 \quad \text{if} \quad \sqrt{I_2} \gtrless \left(\sqrt{I_1} + \sqrt{I_3}\right)/2 \tag{13}$$

Both cases are possible. The case $f_{1\,min} > f_{3\,min}$ will be considered first. Then, diagram 2 can be drawn. It shows schematically for two different values I_0 all three functions $f_\alpha(\lambda)$ in the vicinity of their respective minima. The three upper curves represent the case $I_0 = 0$. According to (13) none of the three polynomials $f_\alpha(\lambda)$ has a real root in the range $I_3 < \lambda < I_1$ in this case. Together with di-

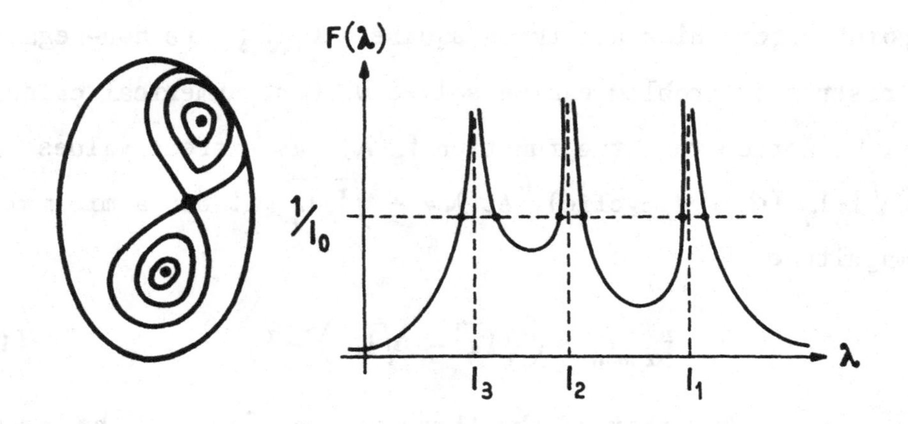

Fig. 1. Fig. 2.

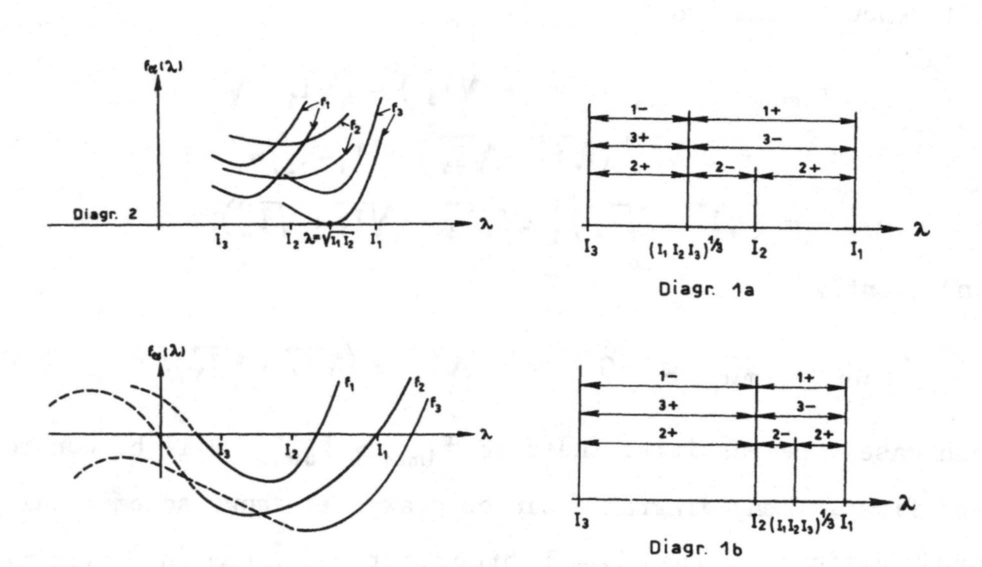

Fig. 3.

agrams 1a,b this means that there is no value λ for which all three squares $u^2_{1,2,3}$ are non-negative. Consequently, there is no boundary line on the sphere for $I_0 = 0$. This result is in accordance with statements made earlier. Diagram 2 shows that with increasing I_0 the situation changes when the minimum $f_{3\,min}$ becomes zero (see the second set of three curves). This happens when

$$I_0 = I_{0\,min} = \left(\sqrt{I_1} - \sqrt{I_2}\right)^2 \tag{14}$$

Then, $f_3(\lambda)$ is zero at $\lambda = \sqrt{I_1 I_2}$ and positive elsewhere in the range $I_3 < \lambda < I_1$ and $f_1(\lambda)$ and $f_2(\lambda)$ are positive everywhere in this range. Comparison with diagrams 1a, b shows that now $\lambda = \sqrt{I_1 I_2}$ is the only value in the range $I_3 < \lambda < I_1$ for which all three squares $u^2_{1,2,3}$ are non-negative. More precisely, (9) yields for $\lambda = \sqrt{I_1 I_2}$ with $I_{0\,min}$ according to (14) the result

$$u_1 = \pm\sqrt{\frac{\sqrt{I_1}}{\sqrt{I_1} + \sqrt{I_2}}}\ , \quad u_2 = \pm\sqrt{1 - u_1^2}\ , \quad u_3 = 0 \tag{15}$$

This is a rotor in the plane of the principal axes of maximum and intermediate moments of inertia. This result was obtained for the case of the upper inequality sigb in (13). For the lower inequality sign a similar solution is found. This time, the curve $f_1(\lambda)$ in diagram 2 reaches the value zero first. This happens at $\lambda = \sqrt{I_2 I_3}$ for

$$I_{0\,min} = \left(\sqrt{I_2} - \sqrt{I_3}\right)^2 \tag{16}$$

and the corresponding rotor axis direction is given by

$$(17) \qquad u_1 = 0 \,, \quad u_2 = \pm\sqrt{\frac{\sqrt{I_2}}{\sqrt{I_2}+\sqrt{I_3}}} \,, \quad u_3 = \pm\sqrt{1-u_2^2}$$

This is a rotor in the plane of the principal axes of minimum and intermediate moments of inertia. Eqs. (14 to (17) give a complete solution for I_{0min}.

For the determination of I_{0max} a somewhat different method must be used. Diagrams 1 a, b show that $\lambda = I_1$, $\lambda = I_2$, $\lambda = I_3$ and $\lambda = (I_1 I_2 I_3)^{1/3}$ are the end points of λ-intervals which are significant for the algebraic signs of $u^2_{1,2,3}$, Diagram 2 shows that with increasing value of I_0 all three curves $f_\alpha(\lambda)$ are shifted down and that for sufficiently large values of I_0 none of the polynomials $f_\alpha(\lambda)$ has a root in the range $I_3 \leq \lambda \leq I_1$. Therefore, it is reasonable to pay attention to the particular values of I_0 for which these polynomials have a root $\lambda = I_1$ or $\lambda = I_2$ or $\lambda = I_3$. Denoting by $I_0^{\alpha\mu}$ the value of I_0 that causes $f_\alpha(\lambda)$ to have a root $\lambda = I_\mu$ $(\mu = 1,2,3)$ we deduce from (11)

$$I_0^{\alpha\mu} = I_\beta + I_\gamma - 3I_\mu + I_\mu^3/(I_\beta I_\gamma) \,, \quad \alpha,\mu = 1,2,3 \,, \quad \alpha,\beta,\gamma \text{ cyclic}$$

By putting in succession $\mu = \alpha$, $\mu = \beta$, $\mu = \gamma$ where as before α, β, γ are supposed to be in cyclic order the results are obtained

$$I_0^{\alpha\alpha} = I_\beta + I_\gamma + I_\alpha \left[I_\alpha^2/(I_\beta I_\gamma) - 3 \right]$$

$$I_0^{\alpha\beta} = (I_\beta - I_\gamma)^2/I_\gamma \ , \quad I_0^{\alpha\gamma} = (I_\beta - I_\gamma)^2/I_\beta \tag{18}$$

$$\alpha = 1,2,3 \qquad \alpha,\beta,\gamma \text{ cyclic}$$

Based on the inequalities $I_3 < I_2 < I_1$ a large number of inequalities can be established between the nine quantities (18). Only the following ones will be neede

$I_0^{21} - I_0^{23} > 0$ (19a)

$I_0^{21} - I_0^{12} > 0$ (19b)

$I_0^{23} - I_0^{13} > 0$ (19c)

$I_0^{21} - I_0^{31} > 0$ (19d)

$I_0^{21} - I_0^{32} > 0$ (19e)

The proofs are so simple that they can be omitted. Now let us suppose that $I_0 = I_0^{21} = (I_1 - I_3)^2/I_3$. Then the function $f_2(\lambda)$ has a root at $\lambda = I_1$ (by definition of I_0^{21}) and no other root at $\lambda \geq I_3$ because of (19a). The function $f_1(\lambda)$ has a root at $\lambda > I_2$ because of (19 - b) and no other root at $\lambda \geq I_3$ because of (19c). Finally, $f_3(\lambda)$ has a root $\lambda > I_1$ because of (19d) and no other root for $\lambda \geq I_2$ because of (19e). These essential properties are depicted in Fig. 3. Superposition with the sign distribution shown in diagrams 1 a,b reveals that $\lambda = I_1$ is the only λ-value that allows three non-negative squares $u_{1,2,3}^2$. If I_0 is made a little smaller than I_0^{21} then a finite range of λ -values in the vicinity of $\lambda = I_1$ exists and for $I_0 > I_0^{21}$ no λ -value at all exists. Thus we conclude that

(20) $$I_{0\,max} = I_0^{21} = (I_1 - I_3)^2 / I_3$$

is the maximum value of I_0 for which boundary lines on the unit sphere occur. The corresponding rotor direction is obtained from (9) as

(21) $$u_1 = u_2 = 0\,, \quad u_3 = \pm 1$$

This is a rotor in the direction of the axis of minimum moment of inertia.

This chapter is concluded with some figures showing boundary lines on unit spheres for a gyrostat with $I_3 = 3$, $I_2 = 5$, $I_1 = 7$ and for five different values of I_0. For a better visualization the unit spheres have been given different sizes, namely diameters proportional to $\sqrt{I_0} \sim h$. For this gyrostat we have $\sqrt{I_2} > (\sqrt{I_1} + \sqrt{I_3})/2$. Consequently, $I_{0\,min} = (\sqrt{7} - \sqrt{5})^2 \approx 0\,1678$ and $I_{0\,max} = 16/3$. In Fig. 4a I_0 is slightly larger than $I_{0\,min}$ so that there is already a finite region indicating rotor directions with four real roots of (4), i.e. with four axes of permanent rotation. For $I_0 = 0{,}27$ another region belonging to four real roots of (4) appears (it is the region caused by the roots of $f_1(\lambda)$ in diagram 2). Both regions overlap at $I_0 = 0{,}48$ thus creating a region belonging to two real roots of (4). With further increase of I_0 the picture becomes very complicated. In order to make out the number of real roots of (4) assigned to a particular region one has to start from some region where the number is known from the previous picture and then to add or subtract two every time a boundary line is crossed. If I_0 is still further increased the picture becomes clearer again. At $I_0 = 1,92$ only regions with four and two real roots of (4) are left. The regions with four real roots, finally, disappear at $I_0 > I_{0max}$.

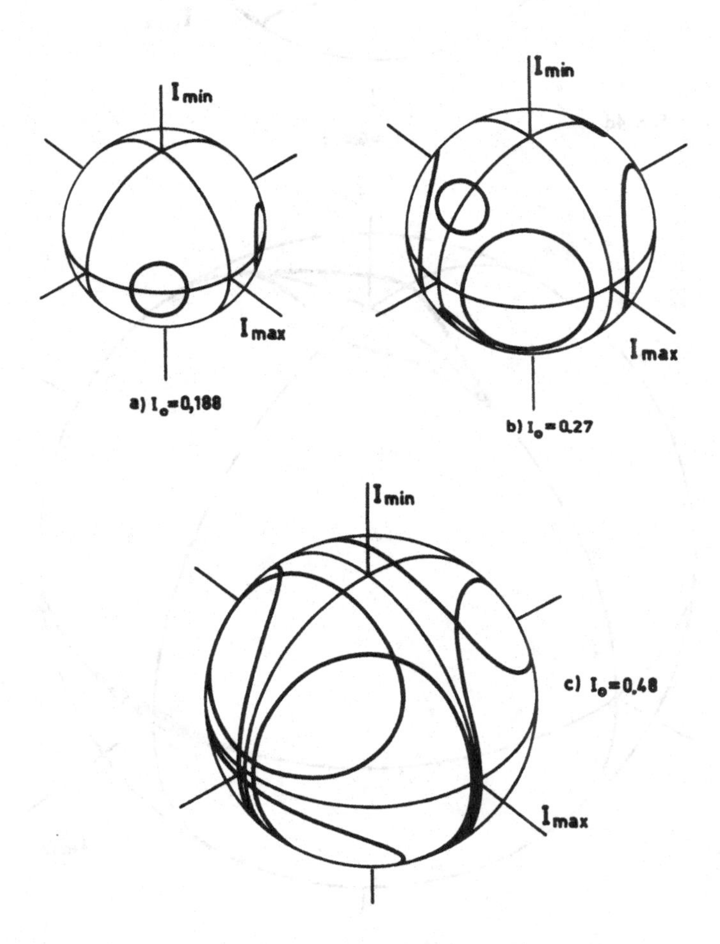

Fig. 4.

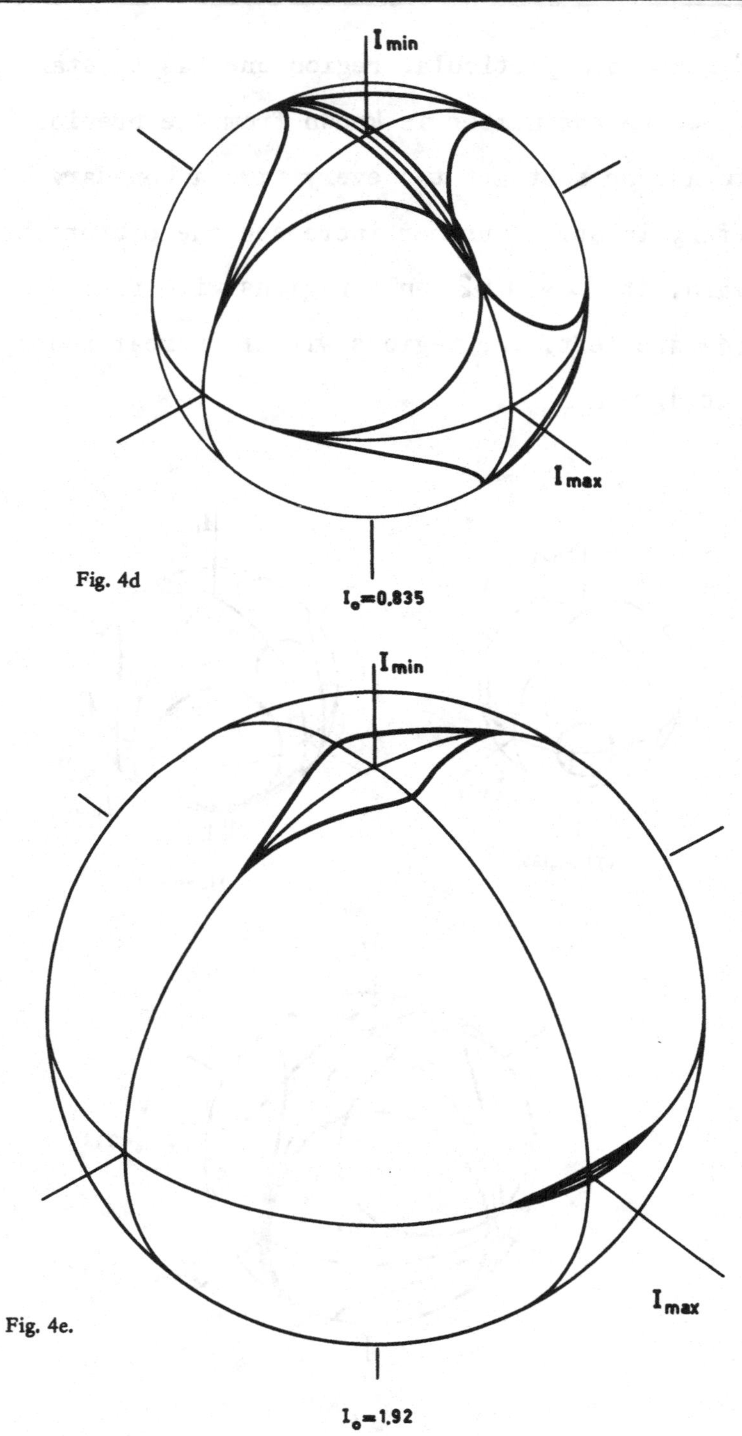

Fig. 4d

Fig. 4e.

7. The General Free Gyrostat (continued)

In this lecture the angular velocity components of a gyrostat are expressed as functions of time. Again, the case of torque-free motion of an unsymmetrical gyrostat with three non-zero components $h_{1,2,3}$ of relative angular rotor momentum is considered. Then, two integrals of motion are available:

$$\sum_{\alpha=1}^{3} I_\alpha \omega_\alpha^2 = 2T \tag{1}$$

$$\sum_{\alpha=1}^{3} (I_\alpha \omega_\alpha + h_\alpha)^2 = 2DT \tag{2}$$

Hence, of the three differential equations of motion only one is needed to determine together with (1) and (2) $\omega_{1,2,3}$ as functions of time. We select the third one which reads

$$I_3 \dot{\omega}_3 - (I_1 - I_2)\,\omega_1 \omega_2 + \omega_1 h_2 - \omega_2 h_1 = 0 \tag{3}$$

This mathematical problem was solved independently by two scientists late last century. In 1889 Wangerin [1] (mathematician at Halle, Germany) published his solution and in 1898 the Italian Volterra, without knowing Wangerin's paper, discovered a completely different way to solve it. From a purely mathematical point of view Volterra's method is more appealing. His final equations for $\omega_{1,2,3}(t)$ display a beautiful symmetry which reflects the cyclical symmetry of the differential equations of motion. Furthermore, only Volterra succeded in integrating also the kinematical (Poisson) differential equations in addition to the Eqs. (1), (2), (3) so that the attitude of the gyrostat in inertial space is given explicitely as a function of time. Volterra's formalism, however, has properties which render it rather inconvenient for applications where numerical calculations of the functions $\omega_{1,2,3}(t)$ are needed. This is due to the fact that these fucntions are expressed in terms of complex variables and that these expressions are so complicated that an analytical separation of real and imaginary parts cannot be achieved (the main problem is presented by a complex modulus of elliptic functions). In this respect Wangerin's approach is simpler because his formalism employs only real quantities. For this reason only Wangerin's method will be presented here.

Multiplication of (1) with some as yet unspecified factor λ and subtraction from (2) results in

$$\sum_{\alpha=1}^{3} I_\alpha (I_\alpha - \lambda) \left(\omega_\alpha + \frac{h_\alpha}{I_\alpha - \lambda}\right)^2 = \lambda \left[\sum_{\alpha=1} \frac{h_\alpha^2}{I_\alpha - \lambda} - 2T\right] + 2DT =$$

$$= -f(\lambda) + 2DT. \tag{4}$$

This function $f(\lambda)$ has simple poles at $\lambda = I_\alpha$ $(\alpha = 1,2,3)$. Therefore, the equation

$$f(\lambda) = 2DT \tag{5}$$

which can be written as quartic equation has at least one real root in each of the ranges $I_3 < \lambda < I_2$ and $I_2 < \lambda < I_1$. Let the root (or one of the roots) in the range $I_3 < \lambda < I_2$ be denoted λ_0 and used as factor λ in (4). Then, (4) becomes

$$\sum_{\alpha=1}^{3} I_\alpha (I_\alpha - \lambda_0) \left(\omega_\alpha + \frac{h_\alpha}{I_\alpha - \lambda_0}\right)^2 = 0$$

and with the substitution

$$w_\alpha = \omega_\alpha + \frac{h_\alpha}{I_\alpha - \lambda_0} , \quad \alpha = 1,2,3$$

$$I_1 (I_1 - \lambda_0) w_1^2 + I_2 (I_2 - \lambda_0) w_2^2 = I_3 (\lambda_0 - I_3) w_3^2 \tag{6}$$

or finally

$$\frac{w_1^2}{(k_1 w_3)^2} + \frac{w_2^2}{(k_2 w_3)^2} = 1 \tag{7}$$

where the quantities

$$k_1 = + \sqrt{\frac{I_3(\lambda_0 - I_3)}{I_1(I_1 - \lambda_0)}} , \quad k_2 = + \sqrt{\frac{I_3(\lambda_0 - I_3)}{I_2(I_2 - \lambda_0)}} \tag{8}$$

have been defined which are real because of the inequalities $I_3 < \lambda_0 < I_2 < I_1$. By (7) an elliptic cone is described (Fig. 1) on the surface of which lies the polhode defined by (1) and (2), therefore, (1) and (7) can be used. Eq. (7) is satisfied by the equations

$$w_1 = k_1 w_3 \sin \upsilon \tag{9}$$

$$w_2 = k_2 w_3 \cos \upsilon \tag{10}$$

The variable υ thus introduced is interpreted as an angle about the axis w_3 (s. Fig. 1). Eqs. (9) and (10) are a parameter representation of the cone. Substitution of (9), (10) and (8) into (1) produces a quadratic equation for w_3 in terms of υ of the form

(11) $$a_1(\upsilon)w_3^2 - 2a_2(\upsilon)w_3 + a_3 = 0$$

where

(12) $$a_1(\upsilon) = I_3\left(\frac{\lambda_0 - I_3}{I_1 - \lambda_0}\sin^2\upsilon + \frac{\lambda_0 - I_0}{I_2 - \lambda_0}\cos^2\upsilon + 1\right)$$

(13) $$a_2(\upsilon) = \frac{h_1 I_1 k_1}{I_1 - \lambda_0}\sin\upsilon + \frac{h_2 I_2 k_2}{I_2 - \lambda_0}\cos\upsilon + \frac{h_3 I_3}{I_3 - \lambda_0} =$$

$$= \sqrt{I_3(\lambda_0 - I_3)}\left(\sqrt{\frac{h_1^2 I_1}{(I_1-\lambda_0)^3}}\sin\upsilon + \sqrt{\frac{h_2^2 I_2}{(I_2-\lambda_0)^3}}\cos\upsilon - \sqrt{\frac{-h_3^2 I_3}{(I_3-\lambda_0)^3}}\right)$$

(14) $$a_3 = \sum_{\alpha=1}^{3}\frac{h_\alpha^2 I_\alpha}{(I_\alpha - \lambda_0)^2} - 2T$$

Later the inequality will be used

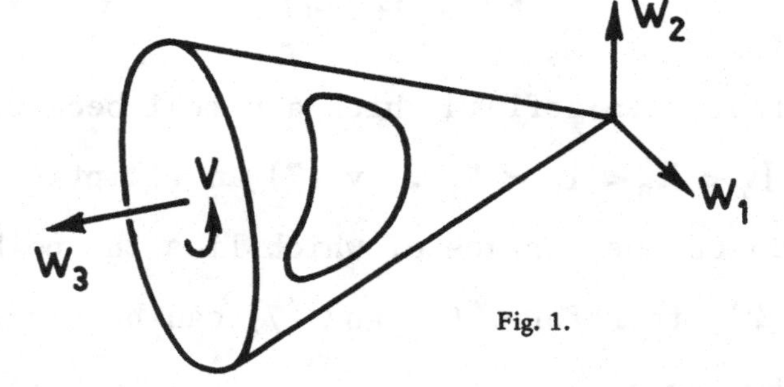

Fig. 1.

(15) $$a_1(\upsilon) \geq \frac{I_1 - I_3}{I_1 - \lambda_0}\, I_3 > 0$$

Eq. (11) has the solution

$$w_3(\upsilon) = \frac{1}{a_1}\left(a_2 \pm \sqrt{a_2^2 - a_1 a_3}\right) \tag{16}$$

With (9), (10), (16) and (6) all three variables $\omega_{1,2,3}$ are expressed in terms of υ. There remains the relationship $\upsilon = \upsilon(t)$ to be established. This is achieved by (3). Substitution of (6) into (3) yields

$$\dot{w}_3 = \frac{1}{I_3}\left[(I_1 - I_2)\left(w_1 - \frac{h_1}{I_1-\lambda_0}\right)\left(w_2 - \frac{h_2}{I_2-\lambda_0}\right) + \right.$$
$$\left. + \left(w_2 - \frac{h_2}{I_2-\lambda_0}\right)h_1 - \left(w_1 - \frac{h_1}{I_1-\lambda_0}\right)h_2\right]$$

For w_1 and w_2 (9) and (10) are substituted. This leads to

$$\dot{w}_3 = \frac{1}{I_3}\left\{w_3^2 (I_1 - I_2) k_1 k_2 \sin\upsilon \cos\upsilon + w_3\left[-\frac{h_2 k_1 (I_1-I_2)}{I_2-\lambda_0}\sin\upsilon - \right.\right.$$
$$\left. - \frac{h_1 k_2 (I_1 - I_2)}{I_1-\lambda_0}\cos\upsilon + h_1 k_2 \cos\upsilon - h_2 k_1 \sin\upsilon\right] +$$
$$\left. + \left[\frac{I_1 - I_2}{(I_1-\lambda_0)(I_2-\lambda_0)} - \frac{1}{I_2-\lambda_0} + \frac{1}{I_1-\lambda_0}\right]h_1 h_2\right\}$$

The last three terms cancel each other. The rest can be rewritten in the form

$$\dot{w}_3 = \frac{w_3}{I_3}\left[\frac{I_3(\lambda_0 - I_3)(I_1 - I_2)}{\sqrt{I_1 I_2 (I_1-\lambda_0)(I_2-\lambda_0)}} w_3 \sin\upsilon \cos\upsilon + \right.$$

$$+\frac{h_1 k_2 (I_2-\lambda_0)}{I_1-\lambda_0}\cos\vartheta - \frac{h_2 k_1 (I_1-\lambda_0)}{(I_2-\lambda_0)}\sin\vartheta\Big] \tag{17}$$

Now (11) is derived with respect to time yielding (with the notation $\dot{} = \frac{d}{dt}$ and $' = \frac{d}{d\vartheta}$)

$$\dot{\vartheta}\, w_3 (a_1' w_3 - 2a_2') = -2\dot{w}_3 (a_1 w_3 - a_2)$$

or with (16)

$$\dot{\vartheta}\, w_3 (a_1' w_3 - 2a_2') = \pm 2\dot{w}_3 \sqrt{a_2^2 - a_1 a_3} \tag{18}$$

Furthermore, with

$$a_1' = -2 I_3 \frac{(\lambda_0 - I_3)(I_1 - I_2)}{(I_1-\lambda_0)(I_2-\lambda_0)} \sin\vartheta \cos\vartheta$$

$$a_2' = \frac{h_1 I_1 k_1}{I_1-\lambda_0}\cos\vartheta - \frac{h_2 I_2 k_2}{I_2-\lambda_0}\sin\vartheta$$

the term in parentheses on the left side of (19) becomes

$$a_1' w_3 - 2a_2' = -2 w_3 I_3 \frac{(\lambda_0 - I_3)(I_1 - I_2)}{(I_1-\lambda_0)(I_2-\lambda_0)} \sin\vartheta\cos\vartheta -$$

$$-2\left(\frac{h_1 I_1 k_1}{I_1-\lambda_0}\cos\vartheta - \frac{h_2 I_2 k_2}{I_2-\lambda_0}\sin\vartheta\right)$$

or substituting (8)

$$a_1' w_3 - 2a_2' = -2\sqrt{\frac{I_1 I_2}{(I_1-\lambda_0)(I_2-\lambda_0)}}\left[\frac{I_3(\lambda_0-I_3)(I_1-I_2)}{\sqrt{I_1 I_2 (I_1-\lambda_0)(I_2-\lambda_0)}} w_3 \sin v \cos v + \right.$$
$$\left. + \frac{h_1 k_2 (I_2-\lambda_0)}{I_1-\lambda_0}\cos v - \frac{h_2 k_1 (I_1-\lambda_0)}{I_2-\lambda_0}\sin v\right]$$

The square bracket in this equation is identical with the one in (17) from which follows

$$w_3 (a_1' w_3 - 2a_2') = -2 I_3 \dot{w}_3 \sqrt{\frac{I_1 I_2}{(I_1-\lambda_0)(I_2-\lambda_0)}}$$

and finally with (18)

$$\dot{v} I_3 \sqrt{\frac{I_1 I_2}{(I_1-\lambda_0)(I_2-\lambda_0)}} = \pm \sqrt{a_2^2 - a_1 a_3} \tag{19}$$

where a_1, a_2, a_3 have to be substituted from (12), (13), (14). This is a differential equation for the function $v(t)$ which in view of the special form of its right side leads to elliptic functions. Wangerin was satisfied with this discovery and closed his paper with the remark that for the actual integration of (19) substitutions proposed by Jacobi in a paper of 1835 have to be used. This last remark is wrong since as will be seen later the radicand in (19), in general, is not positive definite. This, however, is a requirement for the applicability of Jacobi's substitutions.

In the following Wangerin's formalism will be further developed to the point where numerical results are obtainable for $\omega_{1,2,3}(t)$ for arbitrary system parameters and arbitrary initial conditions. It will be seen that in addition to the integration of (19) some problems have to be solved that Wangerin was not aware of. Furthermore, interesting relationships with the osculation problem (not known to Wangerin either) will be established.

By substituting (12), (13), and (14) into (19) after some manipulations the equation is obtained

$$t-t_0 = \pm\sqrt{\frac{-I_1 I_2 I_3}{(I_1-\lambda_0)(I_2-\lambda_0)(I_3-\lambda_0)}}\int_{\upsilon_0}^{\upsilon}(c_1\sin^2\upsilon + c_2\cos^2\upsilon + c_3 +$$

$$+ 2c_4\sin\upsilon + 2c_5\cos\upsilon + 2c_6\sin\upsilon\cos\upsilon)^{-\frac{1}{2}}dt \tag{20}$$

with

$$\begin{cases} c_1 = -\dfrac{a_3-F_1}{I_1-\lambda_0}, & c_4 = -\sqrt{\dfrac{F_3F_1}{(\lambda_0-I_3)(I_1-\lambda_0)}} \\ c_2 = -\dfrac{a_3-F_2}{I_2-\lambda_0}, & c_5 = -\sqrt{\dfrac{F_2F_3}{(I_2-\lambda_0)(\lambda_0-I_3)}} \\ c_3 = +\dfrac{a_3-F_3}{I_3-\lambda_0}, & c_6 = +\sqrt{\dfrac{F_1F_2}{(I_1-\lambda_0)(I_2-\lambda_0)}} \end{cases} \tag{21}$$

where $F_{1,2,3}$ are defined by

$$F_\alpha = \frac{h_\alpha^2 I_\alpha}{(I_\alpha-\lambda_0)}, \quad \alpha = 1,2,3 \tag{22}$$

and $a_3 = F_1 + F_2 + F_2 - 2T$ is the quantity given by (14). For a complete and detailed solution of the elliptic integral (20) there is not enough space here. The reader is referred to the literature on elliptic functions (e.g. 3 , 4 , 5) and to 6 . Omitting the discussion of degenerate cases the solution is achieved as follows. The substitution

$$x = tg\frac{v}{2} \tag{23}$$

yields

$$J = \int_{v_0}^{v}\left(c_1 \sin^2 v + \ldots + 2c_6 \sin v \cos v\right)^{-1/2} dv =$$
$$= 2\int_{x_0}^{x}\left(p_4 x^4 + p_3 x^3 + p_2 x^2 + p_1 x + p_0\right) dx \tag{24}$$

where

$$\left.\begin{array}{ll} p_4 = c_2 + c_3 - 2c_5 \ , & p_0 = c_2 + c_3 + 2c_5 \\ p_3 = 4(c_4 - c_6) & p_1 = 4(c_4 + c_6) \\ p_2 = 2(2c_1 - c_2 + c_3) & \end{array}\right\} \tag{25}$$

Omitting the possible case $p_4 = 0$ the polynomial in x has four roots. Omitting also the case of multiple roots which leads to a solution for J in terms of elementary functions one of the following three cases occurs

(a) all roots are real. We call them $x_{1,2,3,4}$ in the order $x_1 < x_2 < x_3 < x_4$,

(b) two roots are real and two are complex. We call them $x_{1,2} = y_1 \pm i z_1 , x_3 , x_4$,

(c) all four roots are complex. We call them $x_{1,2} = y_1 \pm iz_1$, $x_{3,4} = y_2 \pm iz_2$. Next, real quantities λ_1, λ_2 are calculated from the equations:

case (a)

$$\lambda_{1,2} = \frac{2}{(x_3-x_4)^2}\Big[(x_1+x_2)(x_3+x_4)/2 - x_1x_2 - x_3x_4 \pm \pm\sqrt{(x_1-x_3)(x_2-x_3)(x_1-x_4)(x_2-x_4)}\Big]$$

case (b)

$$\lambda_{1,2} = \frac{2}{(x_3-x_4)^2}\Big[y_1(x_3+x_4) - y_1^2 - z_1^2 - x_3x_4 \pm \pm\sqrt{[y_1(x_3+x_4) - y_1^2 - z_1^2 - x_3x_4]^2 + z_1^2(x_3-x_4)^2}\Big]$$

case (c)

$$\lambda_{1,2} = \frac{1}{2z_2^2}\Big[(y_1-y_2)^2 + z_1^2 + z_2^2 \pm \sqrt{[(y_1-y_2)^2+z_1^2+z_2^2]^2 + [2z_2(y_1-y_2)]^2}\Big]$$

Then real quantities α and β are calculated from the equations

$$\alpha = (b_1 - \lambda_1 b_2)/(\lambda_1 - 1)\,, \quad \beta = (b_1 - \lambda_2 b_2)/(\lambda_2 - 1) \tag{26}$$

where

$$b_1 = \begin{cases} -(x_1+x_2)/2 \text{ in case (a)} \\ -y_1 \quad \text{in case (b) (c)} \end{cases} \qquad b_2 = \begin{cases} -(x_3+x_4)/2 \text{ in case (a)(b)} \\ -y_2 \text{ in case (c)} \end{cases}$$

Finally, real quantities A_1, B_1, A_2, B_2 are calculated form the equations

$$\left.\begin{aligned} A_2 &= (1-\lambda_1)\,/\,(\lambda_2-\lambda_1)\ ; & B_2 &= -(1-\lambda)/(\lambda_2-\lambda_1) \\ A_1 &= \lambda_2 A_2 \ ; & B_1 &= \lambda_1 B_2 \end{aligned}\right\} \quad (27)$$

With these the fourth-order polynomial under the integral in (24) takes the form

$$p_4 x^4 + \ldots + p_0 =$$

$$= p_4\left[A_1(x-\alpha)^2 + B_1(x-\beta)^2\right]\left[A_2(x-\alpha)^2 + B_2(x-\beta)^2\right] \quad (28)$$

The substitution

$$z = \frac{x-\alpha}{x-\beta} \quad (29)$$

then gives J the form

$$J = \frac{2}{(\alpha-\beta)\sqrt{|p_4|}} \quad \left[(a_1 z^2 + b_1)(a_2 z^2 + b_2)\right]^{-1/2} dz \quad (30)$$

where

$$\left.\begin{aligned} a_1 &= A_1 \ \mathrm{sign}\ (p_4)\ ; & a_2 &= A_2 \\ b_1 &= B_1 \ \mathrm{sign}\ (p_4)\ ; & b_2 &= B_2 \end{aligned}\right\} \quad (31)$$

The difference $(\alpha-\beta)$ is equal to zero only if J is not an elliptic integral.i.e. if the polynomial in x has multiple roots. If (26) is neither negative definite nor negative semi-definite then, multiplying all four constants $a_{1,2}$, $b_{1,2}$ by (-1) and/or by interchanging the indices 1 and 2, it can be arranged that one of the six cases of Table 1 applies where by (+) and (−) the algebraic sign of the respective quantities are indicated.

Table 1

case	a_1	b_1	a_2	b_2	$\frac{a_2 b_1 - a_1 b_2}{a_1 b_2}$
1	+	+	+	+	−
2	−	+	+	+	−
3	+	−	+	+	−
4	+	−	+	−	−
5	+	−	−	+	−
6	+	−	−	+	+

Suppose that the necessary changes have been made and that (30) already fits one of the cases of Table 1. Then by the final substitution

$$\varphi = z \left| \frac{a_1}{b_1} \right| \tag{32}$$

a Legendre standard form of elliptic integral is obtained for J which allows to write

$$J = \frac{2}{(\alpha-\beta)\sqrt{|p_4|}} \cdot \frac{1}{C} \left[g(\Phi,k) - g(\Phi_0,k) \right] \tag{33}$$

where Φ_0 and Φ are the limits of the φ-interval of integration, C is a constant, $g(\Phi,k)$ an elliptic integral and k its modulus. The inverse function of (31) is

$$\Phi = \Phi(u,k) \tag{34}$$

where from (33) and (20) for the argument $u = u(t)$ the expression is obtained

$$u(t) = \pm \frac{1}{2}(t-t_0)(\alpha-\beta)C\sqrt{|p_4|}\sqrt{\frac{(I_1-\lambda_0)(I_2-\lambda_0)(I_3-\lambda_0)}{-I_1 I_2 I_3}} +$$

$$+ g(\Phi_0, k). \tag{37}$$

Table 2 summarizes for the six cases of Table 1 the functions $g(\Phi,k)$ and $\Phi(u,k)$, the constant C and the modulus $k = \sqrt{1-k'^2}$.

case	$g(\Phi,k)$	$\Phi(u,k)$	C	k^2
1	$sc^{-1}(\Phi,k)$	$sc(u,k)$	$+\sqrt{a_1 b_2}$	$(a_1b_2-a_2b_1)\ (a_1b_2)$
2	$cn^{-1}(\Phi,k)$	$cn(u,k)$	$-\sqrt{a_2b_1-a_1b_2}$	$a_2b_1\ (a_2b_1-a_1b_2)$
3	$cn^{-1}(1/\Phi,k)$	$nc(u,k)$	$+\sqrt{a_1b_2-a_2b_1}$	$a_1b_2/(a_1b_2-a_2b_1)$
4	$sn^{-1}(\Phi,k)$	$sn(u,k)$	$+\sqrt{-a_1b_2}$	$a_2b_1\ (a_1b_2)$
5	$sn^{-1}\left(\frac{1}{k}\sqrt{1-k'^2\Phi^2},k\right)$	$\frac{1}{k'}dn(u,k)$	$-\sqrt{a_1b_2}$	$(a_1b_2-a_2b_1)\ (a_1b_2)$
6	$dn^{-1}(\Phi,k)$	$dn(u,k)$	$+\sqrt{a_2b_1}$	$(a_2b_1-a_1b_2)\ (a_2b_1)$

Table 2. Glaisher's notation is used. The exponent -1 denotes an inverse function.

Eqs. (32) and (33) together with (30), (27) and (23) define the parameter v as a function of time. In order to illustrate how the formalism developed so far is used let us assume that the parameters $I_{1,2,3}$ and $h_{1,2,3}$ as well as the initial conditions $\omega_{1,2,3}(t=t_0)$ are given and that $\omega_{1,2,3}(t=t_1)$ are to be determined. This requires the calculation of the following quantities:

1) $2T$ and $2DT$ from (1), (2) for the initial conditions
2) $I_3<\lambda_0<I_2$ from the quartic equation (5)
3) k_1, k_2 from (8)

4) a_3 , F_1 , F_2 , F_3 from (14) and (22)

5) $c_1, \ldots, c_6$ from (21)

6) $p_0, \ldots, p_4$ from (25)

7) the roots of the quartic $p_4 x^4 + \ldots + p_0 = 0$ (it is assumed here that it has no multiple roots)

8) α , β , A_1 , B_1 , A_2 , B_2 from (26) and (27)

9) a_1 , b_1 , a_2 , b_2 from (31)

10) new quantities a_1 , b_1 , a_2 , b_2 such that one of the case of Table 1 applies (case N)

11) for case N the modulus k and the constant C from Table 2

12) $w_{1,2,3}(t_0)$ from (6)

13) $v(t_0)$ from (9), (10)

14) $x(t_0)$ from (23)

15) $z(t_0)$ from (29)

16) $\Phi_0 = \varphi(t_0)$ from (32)

17) the incomplete elliptic integral $g(\Phi_0, k)$ for case N of Table 2

18) $u = u(t_1)$ from (35); the proper choice of the algebraic sign will be discussed later

19) $\Phi = \Phi(u,k)$ fro case N of Table 2

20) $z = z(\Phi)$ from (32)

21) $x = x(z)$ from (29)

22) $v = v(x)$ from (23)

23) $a_1(v)$, $a_2(v)$ from (12), (13)

24) two values $w_3(v)$ from (16)

25) two values each for $\mathfrak{w}_1(\mathfrak{v}), \mathfrak{w}_2(\mathfrak{v})$ from (9), (10)

26) two values each for $\omega_1(\mathfrak{w}_1)$, $\omega_2(\mathfrak{w}_2)$, $\omega_3(\mathfrak{w}_3)$ from (6).

This list shows the complexity of the problem. It indicates the numerical difficulties and points out problems which still remain to be solved. Numerical difficulties are caused by the succession of two quartic equations and one quadratic equation (for $\lambda_{1,2}$) the coefficients of each one of which depend on the roots of the respective previous equation. These roots have to be determined with great accuracy. Otherwise the result of step N° 26 above will be meaningless. A completely satisfactory computer subroutine was found in [7] (it is interesting that subroutined based on the exact algebraic formulas for the roots of quartic equations turned out to be unsatisfactory). In steps N° 18 and 26 problems connected with algebraic signs of square roots remain to be solved. An answer to these questions requires a better understanding of the geometry of the polhodes on the cone (7). We start out by an investigation of (5). The derivative with respect to λ of $f(\lambda)$ is

$$\frac{\partial f}{\partial \lambda} = -\left[\sum_{\alpha=1}^{3} \frac{h_\alpha^2 I_\alpha}{(I_\alpha - \lambda)^2} - 2T\right]$$

The expression is familiar from the previous lecture on permanent rotations. The condition for a state of permanent rotation can be written now in the form

(36) $$\frac{\partial f}{\partial \lambda} = -\left[F(\lambda) - 2T\right] = 0$$

From the real roots λ of this sextic equation the corresponding permanent angular velocity components were obtained as

(37) $$\omega_\alpha = \frac{-h_\alpha}{I_\alpha - \lambda} .$$

The relationship between the functions $F(\lambda)$ and $f(\lambda)$ is shown in Fig. 2. The function $f(\lambda)$ with first-order poles $\lambda = I_\alpha$ $(\alpha = 1,2,3)$ and with the asymptote $2T\lambda + h^2$ has extreme values where the osculation condition $F(\lambda) - 2T = 0$ has roots. From this follows that any value λ satisfying (36) is also a double-root of (5) if for D the osculation value D is used which is calculated from (2) with $\omega_{1,2,3}$ substituted from (37). In Fig. 2 the six values $D_1, \ldots, D_6$ are indicated which belong to the six permanent rotations of this particular gyrostat. For motions with an angular momentum integral (2) for D-value in the range $D_2 < D < D_3$ Eq. (5) has three real roots in the range $I_3 < \lambda < I_2$ which can be used as λ_0 in Wangerin's formalism. It will now be shown that for a complete solution of the equations of motion the intermediate one of the three roots must be chosen. One recognizes that the quantity a_3 of (14) is identical with $F(\lambda_0) - 2T$. Hence, it is zero if the motion investigated is one of the permanent rotations belonging to D_2 or D_3 and it is negative for motion with $D_2 < D < D_3$ if the intermediate real root is chosen. From (15) follows that then the function $a_2^2(\upsilon) - a_1(\upsilon)a_3$ is positive independent of υ. Hence, for any value $0 \leq 2\upsilon \leq 2\pi$ Eq. (16) has two different real roots $w_3(\upsilon)$. This means that the cone (7) is wrapped in two completely separated polhodes as shown in Fig. 3. If instead of the intermediate root one of the other two is used as λ_0 then a_3 is positive and the function $a_2^2 - a_1 a_3$ is not positive over the full range $0 \leq \upsilon \leq 2\pi$. This means that there is only one polhode on the cone as shown in Fig. 1. The other one of the two shown in Fig. 3 is lost. Thus we conclude: If D has a value such that (5) has three real roots in the range $I_3 < \lambda < I_2$ then the intermediate one must be used. Fig. 2 suggests that the same situation arises in the case of motions with $D_4 < D < D_5$. Then (5) has three real roots in the range $I_2 < \lambda < I_1$. Indeed, this is true. Suppose that instead of $I_3 < \lambda_0 < I_2$ the root (or one of the roots)of (5) in the range $I_2 < \lambda_0 < I_1$ is used as λ_0. Then the formalism developed earlier is unchanged up to Eq. (6). The Eqs. (7) through (10), however, have to be changed if only relationships between real quantities are desired. They are replaced by

$$\frac{w_2^2}{(k_2 w_1)^2} + \frac{w_3^2}{(k_3 w_1)^2} = 1\,, \quad k_2 = +\sqrt{\frac{I_1(I_1-\lambda_0)}{I_2(\lambda_0-I_2)}}$$

$$k_3 = +\sqrt{\frac{I_1(I_1-\lambda_0)}{I_3(\lambda_0-I_3)}}$$

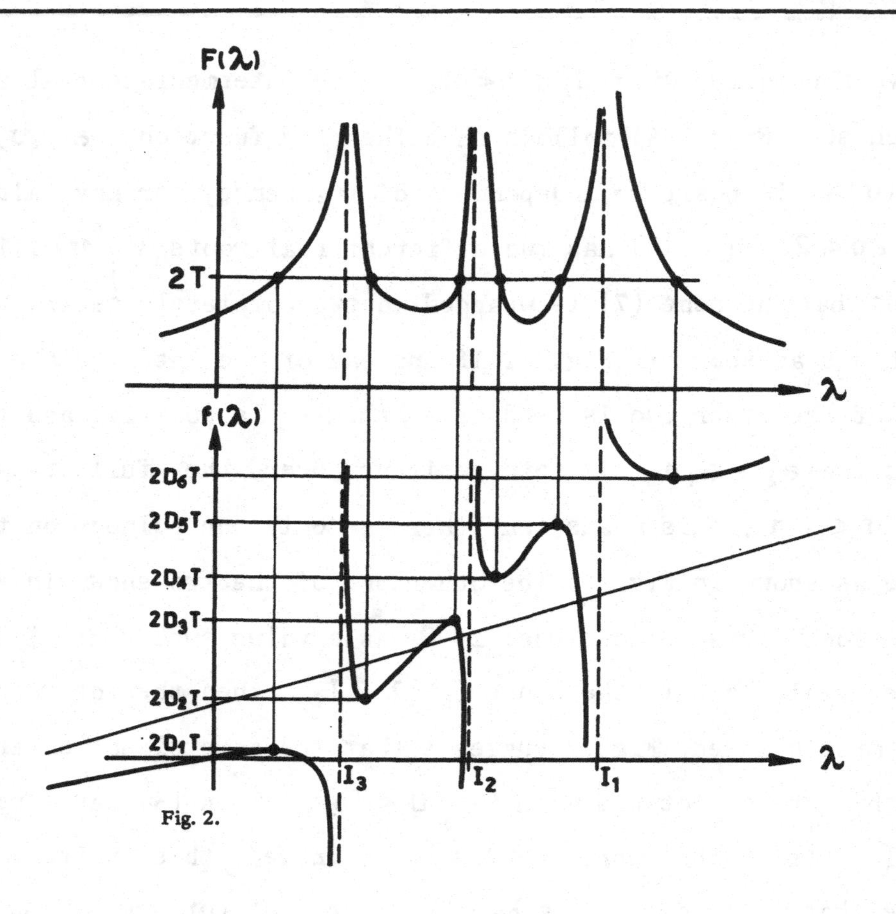

Fig. 2.

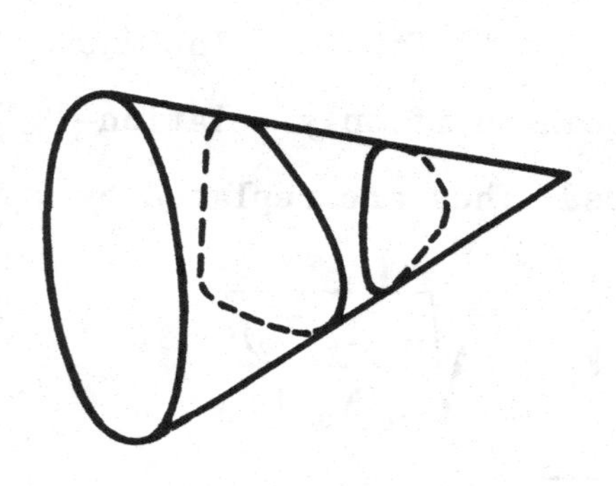

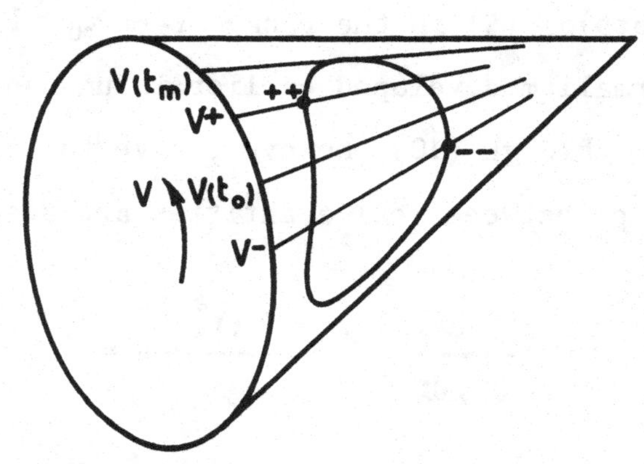

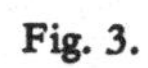

Fig. 3.

Fig. 4.

$$w_2 = k_2 w_1 \sin \vartheta$$

$$w_3 = k_3 w_1 \cos \vartheta$$

Starting from these equations all the steps are carried out that furnished the differential equation (19) for $\vartheta(t)$. Comparison of the equations (6) through (19) with the corresponding ones obtained now shows that the new equations can be obtained from the original ones by raising in cyclic fashion the indices of $I_{1,2,3}$, $h_{1,2,3}$, $\omega_{1,2,3}$ and $k_{1,2}$ by 1 (1 is replaced by 2, 2 by 3 and 3 by 1). Thus, by a very simple maneuver a computer program written for the case $I_3 < \lambda_0 < I_2$ can be adapted to the case $I_2 < \lambda_0 < I_1$. One changes the indices of I_α, h_α and $\omega_\alpha(t_0)$ in the described manner (after solving the equation $f(\lambda) = 2DT$ for λ_0 !). One then carries out the entire program. When the final results for $\omega_{1,2,3}$ are obtained one changes back the indices. If there are three real roots of (5) in the range $I_2 < \lambda < I_1$ then, as before, the intermediate one has to be used.

We now turn to the question of how to select the algebraic signs of the square roots in (16) and (20). From the mathematical development leading to Eq. (20) one concludes that the plus sign in (20) belongs to the plus sign in (16) and likewise the minus sign in (20) to the minus sign in (16). This is illustrated in Fig. 4 which shows the cone (7) for the case of a polhode with one single closed curve. For some value $t = t_1 \neq t_0$ two values u^+ and u^- are calculated from (35). They determine two corresponding values v^+ and v^- furnishing two generating lines on the cone which intersect the polhode in four points (Fig. 4). Only two of these four points are significant, namely the one for which the algebraic signs in (16) and (35) are the same. They are denoted ++ and -- in Fig. 4. If for the plus sign in (35) v increases with increasing time t then it decreases for the minus sign. The effect on the points ++ and -- is shown by arrows which indicate that the sense in which a polhode is travelled through is independent of the choice of algebraic sign to be used in (35) can now be determined by calculating for $t = t_0$, i.e. for $u^+ = u^- = g(\Phi_0, k)$ the corresponding sets of values $\omega_{1,2,3}(t_0)$. Only one of them fits the given initial conditions. This gives us the algebraic sign to be used in (16). The same algebraic sign is used in (35) for all times t . Fig. 4 shows, however, that with this not all problems are solved. Suppose that starting at $t = t_0$ the algebraic sign thus determined causes $v(t)$ to increase with increasing t . At some time $t = t_m$ the elliptic function $v(t)$ reaches a maximum for which the two roots $w_3(v(t_m))$ of (16) coincide as shown in Fig. 4. For $t > t_m$ $v(t)$ decreases again. Obviously, wrong results are obtained for $\omega_{1,2,3}(t)$ if for $t > t_m$ the same algebraic sign in (35) (a switch here would cause discontinuities) and to change the one in (16) every time a double root w_3 occurs in (16). This happens when the radicand $a_2^2 - a_1 a_3$ has a double root. In the course of automatic calculations such double roots are difficult to detect. Therefore, the fact is used that the expression $a_2^2 - a_1 a_3$ also appears in (19) and that it was expressed there by new variables x, z and φ. The expression in terms of φ was said to yield Legendre's standard forms of elliptic integrals. In the case N° 2 of Table 1, for instance, the radicand has the form $\text{const}\,(1-\varphi^2)\,(k'^2 + k^2\varphi^2)$. For Eq. (16) this means that the same sequence of substitutions would lead to the expression $a_2^2 - a_1 a_3 = \text{const}\,(1-\Phi^2)\,(k'^2+k^2\Phi^2)$ where now Φ is the function $\Phi(u,k)$ from Table 2. Thus, in the case of N° 2 of Table 2 one can write

$$a_2^2 - a_1 a_3 = \text{const}\,(1-\text{cn}^2 u)(k'^2+k^2\text{cn}^2 u) = \text{const}\cdot\text{sn}^2 u\,(k'^2+k^2\text{cn}^2 u).$$

Hence, $a_2^2 - a_1 a_3$ has double roots where $\text{sn}\,u$ has single roots. They can easily be detected automatically. The same approach is applicable to all other cases of Table 2. The results are listed in Table 3.

case	1	2	3	4	5	6
$h(u)$	no double root	sn u	sn u	cn u	sn u cn u	sn u cn u

Table 3

We are now in a position to fully answer the questions concerning algebraic signs in steps N° 18 and 26 of the calculations described earlier. The sign to be used in (35) is determined by a test calculation of $\omega_{1,2,3}(t_0)$. The sign to be used in (16) for $t = t_1$ is found by first calculating the number of sign changes of the function $h(u)$ of Table 3 between $u(t_0)$ and $u(t_1)$. If this number is even then the same sign is used as in (35). If it is odd the opposite sign is used.

The result obtained so far enable one to write a computer program which automatically calculates $\omega_{1,2,3}(t)$ for arbitrary parameters $I_{1,2,3}$ (non-symmetric) and $h_{1,2,3}$ (non-zero), for arbitrary initial conditions and for arbitrary values of time t. This program, however will fail in the case where the integral (20) is not an elliptic integral. It was shown that for permanent rotations a root of (36) is, simultaneously, a double root λ_0 of (5) and that a_3 is zero. Eqs (6) and (37) show that then $w_1 = w_2 = w_3 = 0$ is a solution. It is the solution of (16) belonging to the minus sign, and it is represented by the apex of the cone (7). The solution belonging to the plus sign of (16) is found as follows. For $a_3 = 0$ Eq. (20) in view of (19) and (13) is expressed by a non-elliptic integral in the form

$$t - t_0 = \pm\sqrt{\frac{-I_1 I_2 I_3}{(I_1-\lambda_0)(I_2-\lambda_0)(I_3-\lambda_0)}} \int_{v_0}^{v} \frac{dv}{s_1 \sin v + s_2 \cos v + s_3}$$

with

$$s_1 = +\sqrt{\frac{h_2^2 I_1}{(I_1-\lambda_0)^3}} \qquad s_2 = +\sqrt{\frac{h_2^2 I_2}{(I_2-\lambda_0)^3}} \qquad s_3 = -\sqrt{\frac{-h_3^2 I_3}{(I_3-\lambda_0)^3}}$$

The integral alone has the solution

$$\frac{}{s_1 \sin v + s_2 \cos v + s_3} = \begin{cases} \dfrac{2}{\sqrt{\Delta}} \tan^{-1} \dfrac{s_1 + (s_3 - s_2) \tan v/2}{\sqrt{\Delta}} & \text{for } \Delta > 0 \\ \dfrac{-2}{\sqrt{\Delta}} \tanh^{-1} \dfrac{s_1 + (s_3 - s_2) \tan v/2}{\sqrt{-\Delta}} & \text{for } \Delta < 0 \\ \dfrac{-2}{s_1 + (s_3 - s_2) \tan v/2} & \text{for } \Delta = 0 \end{cases} \tag{38}$$

which depends on the algebraic sign of

$$\Delta = s_3^2 - s_2^2 - s_1^2 = - \sum_{\alpha=1}^{3} \frac{h_\alpha^2 I_\alpha}{(I_\alpha - \lambda_0)^3} = \frac{1}{2} \left. \frac{\partial F}{\partial \lambda} \right|_{\lambda = \lambda_0} \tag{39}$$

From (38) it is seen that for $\Delta > 0$ $v(t)$ is a specific function and, hence, the permanent rotation is stable. For $\Delta < 0$ $v(t)$ is aperiodic, and the permanent rotation is unstable. Eq. (39) shows that $\Delta > 0$ belongs to D_2 and $\Delta < 0$ to D_3 of Fig. 2. The case $\Delta = 0$ is given when (36) has a double root. This was shown in lecture N° 6 to be the critical case of transition from six to four or from four to two axes of permanent rotation. The inverse function of (38) for $\Delta = 0$ is of the type $\mathrm{tg}(v/2) = = \mathrm{const} + \mathrm{const}/(t - t_0)$ which indictes a non-periodic solution. It was shown above that if the osculation condition (36) has two separate roots $I_3 < \lambda < I_2$ then the smaller one yields a stable and the larger one an unstable permanent rotation. If there are two separate real roots in the range $I_2 < \lambda < I_1$ then the situa-

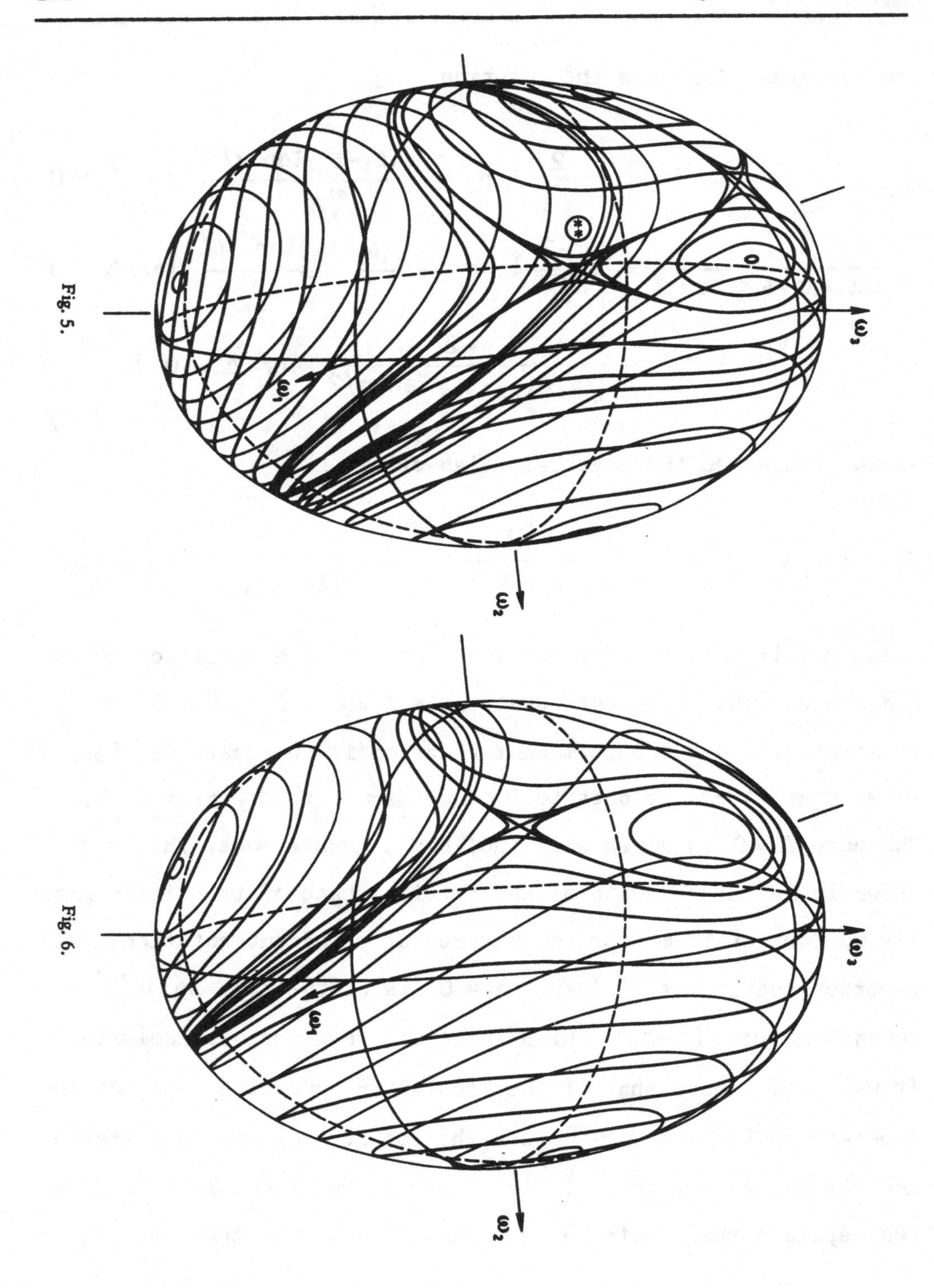

Fig. 5.

Fig. 6.

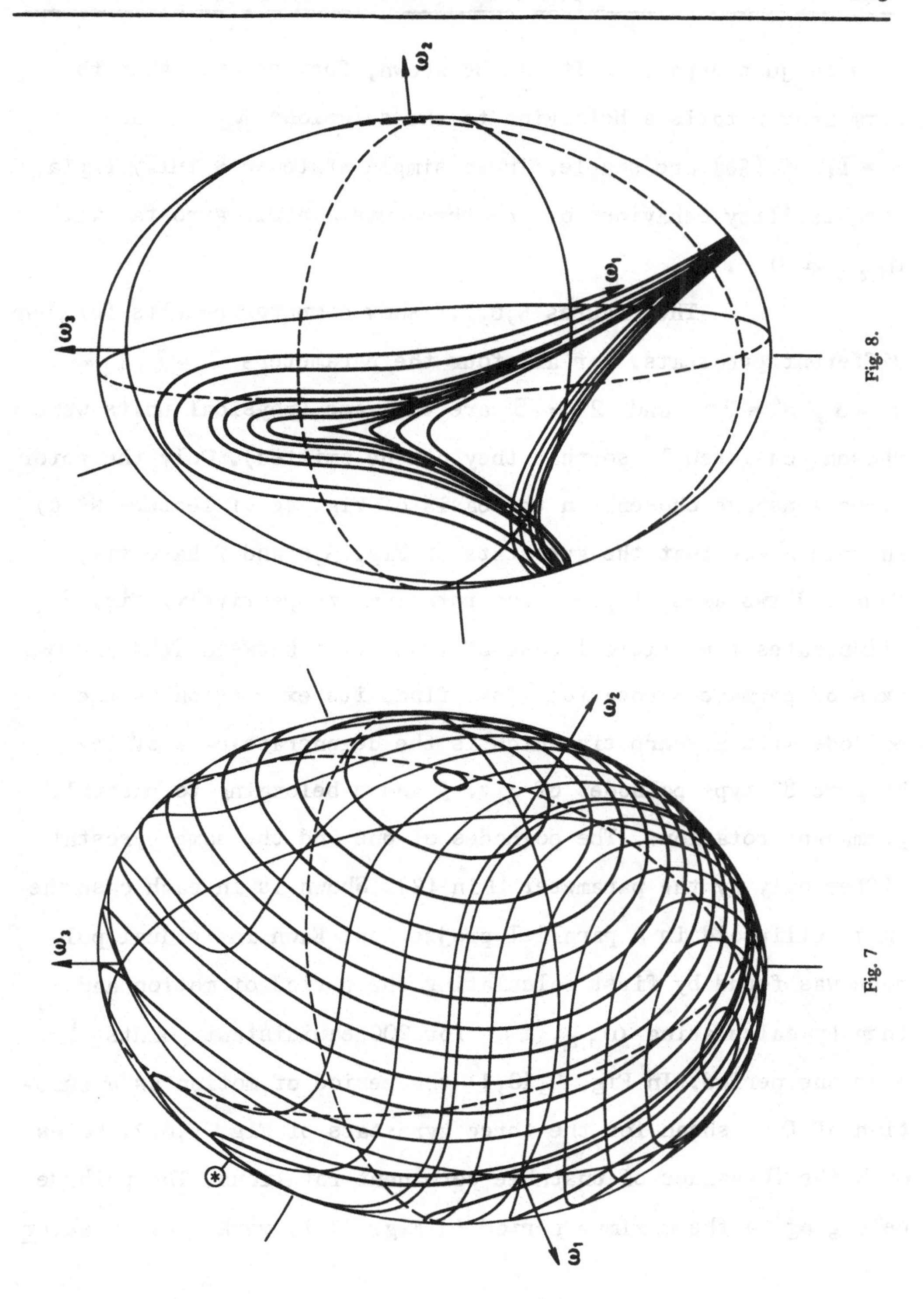

Fig. 8.

Fig. 7

tion is just opposite. It can be shown, furthermore, that the permanent rotations belonging to the solutions $\lambda > I_3$ and $\lambda > I_1$ of (36) are stable. These simple statements fully explain the stability behaviour of the three unsymmetric gyrostat with $h_{1,2,3} \neq 0$.

The figures 5,6,7,8 show computer results for four different gyrostats. For all four the parameters $I_1 = 7$, $I_2 = I_3 = 3$, $h^2 = 36$ and $2T = 75$ are the same (physical units were chosen consistently so that they can be omitted). Only the rotor directions are chosen (on the basis of Fig. 4c of lecture N° 6) in such a way that the gyrostats of Fig. 5,6 and 7 have six, four and two axes of permanent rotation, respectively. Fig. 8 illustrates the critical case of transition between four and two axes of permanent rotation. This finds its expression in the polhode with a sharp tip which is the degenerate case of the "figure 8" type polhodes of Fig. 5 and 6 belonging to unstable permanent rotations. The polhodes of one and the same gyrostat differ only by the parameter D in (2). Shown is in each case the enrgy ellipsoid in a parallel projection. Each individual polhode was found by first calculating the period of motion and then by calculating $\omega_{1,2,3}(t)$ for 200 equidistant points t over one period. In Fig. 9,10,11 the period of motion as a function of D is shoan for the three gyrostats of Fig. 5,6,7. Poles mark the D-values of unstable permanent rotations. The polhode belonging to the maximum period in Fig. 11 is marked by an aster

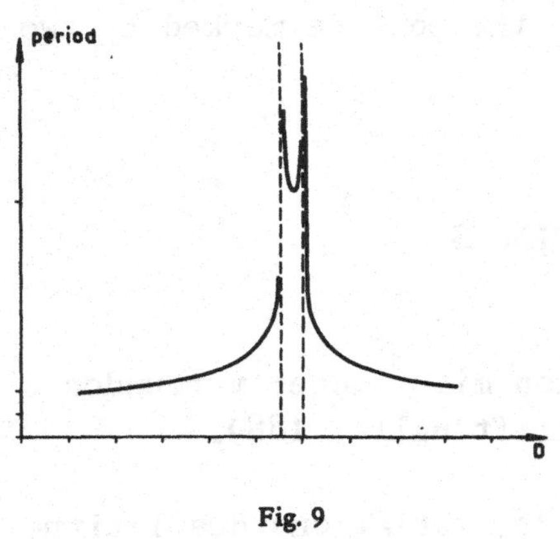

Fig. 9

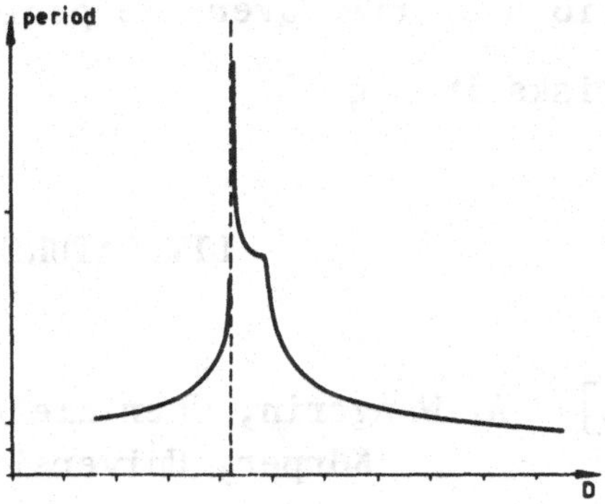

Fig. 10

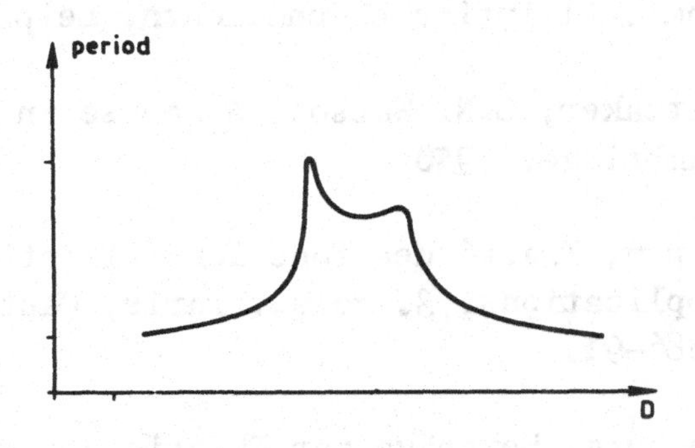

Fig. 11

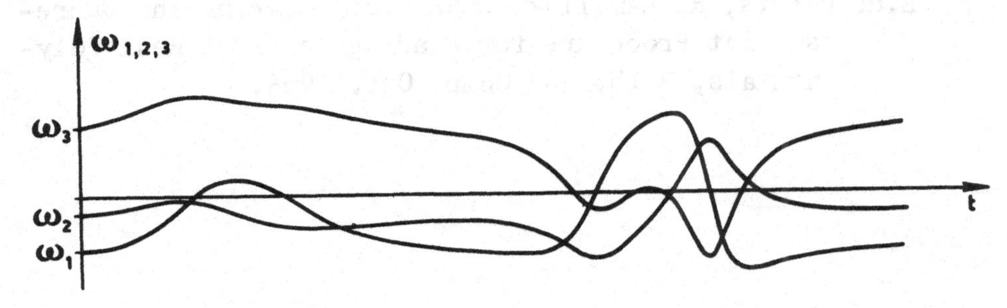

Fig. 12

isk in Fig. 7. In Fig. 12 the components $\omega_{1,2,3}$ are shown as functio,s of time over one period for the polhode marked by two asterisks in Fig. 5.

LITERATURE REFERENCES

[1] A. Wangerin, Über die Rotation miteinander verbundener Körper, Universitätschrift Halle, 1889.

[2] V. Volterra, Sur la théorie des variations des latitudes, Acta Math. 22, 1898.

[3] F. Tricomi, Elliptische Functionen, Leipzig, 1948.

[4] E.T. Whittaker, G.N. Watson, A Course in Modern Analysis, Cambridge, 1950

[5] G.H. Halpen, Traité des fonctions elliptiques et leurs applications, 3. vols., Paris, Gautheir-Villars, 1886-91.

[6] W. Wittenburg, Beiträge zur Theorie des Gyrostaten, Habilitationschrift, to be published.

[7] E.H. Bareis, R. Hamelink, RSSR Root Squaring and Subresultant Procedure for Finding Zeros of Real Polynomials, Math. and Comp. Oct. 1965.

8. Gravitationally Stabilized Rigid Bodies and Gyrostats

The classical libration problem of Lagrange is one forerunner of the analogous problem for the gyrostat. Another is the problem of an orbiting symmetric spinning body ated upon by gravitational torques. The equilibrium orientation of such a body with its spin axis normal to the plane of the (circular) orbit is implicit in several works on spin stabilization; but the first to treat its stability explicitly was Thomson [1] . (His stability phase diagram was incomplete, and was extended by Kane, Marsh and Wilson [2] shortly thereafter.) Likins [3] was the first to demonstrate that a spinning body can have other equilibrium orientations of the spin axis besides the one normal to the orbit plane: specifically, the spin axis can remain in the plane normal to the orbital path but inclined to the orbit normal, or in the plane normal to the local geocentric vertical and again inclined to the orbit normal. These equilibria have precise counterparts in the case of the orbiting gyrostat.

Kane and Mingori treat the simplest case of the orbiting gyrostat in [4] , that in which the rotor is aligned with a principal axis and is aligned with the normal to the orbit plane in the equilibrium orientation. They use a numerical treatment involving Floquet theory to investigate stability. However, Anchev [5] apparently was the first to take up in a quite general way the problem of the equilibria of orbiting gyrostats and their stability. His approach was based on an Eulerian form of the rotational dynamical equations, the postulation of certain basic classes of equilibria, and the determination of the relationships among the various parameters of his problem such that the dynamical equations could be satisfied by the body retaining a fixed orientation in his orbiting reference frame. An important aspect of this problem is that the dynamical basis includes both translational and rotational equations and the coupling that exists between these through the gravitational potenttial of the gyrostat - which is small but not a point mass - and the massive central body. Certain of the results for this "unrestricted problem" differ from those for the "restricted problem" in which the orbital path is assumed to be unaffected by the orientation, but while this difference is conceptually important, it is shown in [6] to be an insignificant difference. This lecture concerns only the restricted problem.

Consider a body orbiting in a circular orbit in an inverse square field. A vector basis $\{\underline{\xi}_\alpha\}$ is defined as follows. (See Fig. 4.) The outward geocentric vertical at a satellite center of mass is denoted $\underline{\xi}_3$. The normal to the instantaneous orbit plane (thus the trajectory binormal) is denoted $\underline{\xi}_2$, taken in a positive sense such that when $\underline{\xi}_1$ is formed as

$\underline{\xi}_2 \times \underline{\xi}_3$ it has the sense of forward motion on the trajectory. It is postulated that the gravitational torque is the only external torque on the body. The existence of this torque was first demonstrated by d'Alambert in 1749, and its forms and applications have a two-century history which is not germain for present purposes. Suffice it that vector-dyadic forms have been given within the last five years by Nidey and Lur'e, and that these have been in common use. Thus it can be considered a familiar result that the vector torque is given by

$$\underline{L} = \frac{3k}{r^3} \underline{\xi}_3 \times \mathbf{I} \cdot \underline{\xi}_3 \tag{1}$$

where k is gravitational constant of the attracting center, r is the radial distance from that center to the satellite, and $\mathbf{I}$ is the inertia dyadic for the body about its center of mass.

The rules given in Lecture 1 permit the vector-dyadic form to be written as a matrix form in body axis. If L, ξ_3 and I are the matrices of components of $\underline{L}$, $\underline{\xi}_3$ and $\mathbf{I}$ respectively,

$$L = 3\,\omega_0^2\, \tilde{\xi}_3\, I\, \xi_3 \tag{2}$$

We have used the fact theat for a circular orbit, k/r^3 is the square of the orbit frequency. The dynamical equation for a gyrostat acted upon by such a torque is

(3) $$I\dot{\omega} + \tilde{\omega}(I\omega + h) = 3\omega_0^2 \tilde{\xi}_3 I \xi_3$$

The existence and uniqueness of stability was established by a new argument by Likins and Roberson [7] for the case of a rigid body $(h = 0)$. It is instructive to review that quickly before proceeding to the gyrostat. First note that a necessary and sufficient condition for the body's equilibrium in the rotating ξ-frame is that the body have the same angular velocity as the frame at all times: that is, $\underline{\omega} = \omega_0 \underline{\xi}_2$. Using this in the dynamical equation gives the equilibrium condition

(4) $$\tilde{\xi}_2 I \xi_2 = 3 \tilde{\xi}_3 I \xi_3$$

where ξ_α is the matrix of components of $\underline{\xi}_\alpha$ in the body frame for $\alpha = 1,2,3$. Any matrix representing a vector in space can be represented as a linear combination of the ξ_α. In particular, we can write

(5) $$\begin{aligned} I\xi_2 &= a\xi_1 + b\xi_2 + c\xi_3 \\ I\xi_3 &= d\xi_1 + e\xi_2 + f\xi_3 \end{aligned}$$

Putting these into Eq. 4, using $\tilde{\xi}_\alpha \xi_\beta = \xi_\gamma$ (α, β, γ cyclic), we get

$$-a\xi_3 + c\xi_1 = 3(d\xi_2 - e\xi_1)$$

which implies $a = d = 0$ and $c = -3e$. But also $\xi_3^T I \xi_2 = \xi_2^T I \xi_3$, so $c = e$. It follows that $c = e = 0$, and we are left with

$$\begin{aligned} I\xi_2 &= b\xi_2 \\ I\xi_3 &= f\xi_3 \end{aligned} \tag{6}$$

In short, both ξ_2 and ξ_3 are eigenvectors of the inertia matrix. (So ξ_1 also must be.) Equilibrium, therefore, is with the principal axes of inertia disposed along the base vectors of the ξ-frame.

The same method was applied by Roberson and Hooker [8] to attempt to find the equilibrium orientations for the gyrostat. The equilibrium condition corresponding to Eq. 4 becomes

$$\tilde{\xi}_2 (I\xi_2 + h/\omega_0) = 3\tilde{\xi}_3 I \xi_3 \tag{7}$$

Now suppose that

$$I\xi_1 = \lambda\xi_1 + a_2\xi_2 + a_3\xi_3 \tag{8a}$$

$$I\xi_2 = b_1\xi_1 + \mu\xi_2 + b_3\xi_3 \tag{8b}$$

$$I\xi_3 = c_1\xi_1 + c_2\xi_2 + \nu\xi_3 \tag{8c}$$

$$h/\omega_0 = J_\alpha \xi_\alpha \tag{8d}$$

Note that h usually is the known quantity in Eq. 8d, the J_α being unknown a priori because it is non known how the body is oriented with respect to the ξ-frame. However, it is helpful to approach the problem by thinking of them as known. Equations 7 and

8 imply $c_1 = 0$, $b_1 = -J_1$, $b_3 = -J_3 - 3c_2$. Next, $\xi_\alpha^T I \xi_\beta = \xi_\beta^T I \xi_\alpha$ imply $a_2 = b_1$, $a_3 = c_1$, $b_3 = c_2$. This leaves

(9a) $$(I - \lambda E)\xi_1 + J_1 \xi_2 = 0$$

(9b) $$J_1 \xi_1 + (I - \mu E)\xi_2 + \frac{1}{4} J_3 \xi_3 = 0$$

(9c) $$\frac{1}{4} J_3 \xi_2 + (I - \nu E)\xi_3 = 0$$

where E is the 3×3 unit matrix, as the equilibrium conditions. Note that J_2 does not appear, as might be anticipated on physical grounds.

Recognize that $\xi_\alpha^T \xi_\alpha = 1$ (α not summed) and $\xi_\alpha^T \xi_\beta = 0$ $(\alpha \neq \beta)$ because of the orthonormality of $\{\underline{\xi}_\lambda\}$. Then the following theorems are obvious, in view of the eigenvector character of a principal axis as regards the inertia matrix:

TH. 1 The condition that $\underline{\xi}_1$ coincide with a principal axis of inertia implies and is implied by $J_1 = 0$.

TH. 2 The condition that $\underline{\xi}_3$ coincide with a principal axis of inertia implies and is implied by $J_3 = 0$.

TH. 3 The condition that $\underline{\xi}_2$ be a principal axis implies and is implied by $J_1 = J_3 = 0$.

The case $J_1 = J_2 = J_3 = 0$ is the classical libration problem which we do not consider further. The case $J_1 = J_3 = 0$, $J_2 \neq 0$ is well defined by TH. 3 and Eqs. 9: all principal axes coincide with $\underline{\xi}_\lambda$ as in the classical problem, the principal axis carrying the internal angular momentum necessarily coincident with the normal to the orbit plane. It can be shōwn by listing exclusive and exhaustive physical situations that the only other cases that need be considered are $J_1 = 0$, $J_3 \neq 0$; $J_1 \neq 0$, $J_3 = 0$; $J_1 \neq 0$, $J_3 \neq 0$. Even though J_α are not "known" parameters, no other physical equilibrium possibilities exist.

We shall consider the case $J_3 = 0$, $J_1 \neq 0$; the case with $J_1 = 0$, $J_3 \neq 0$ can be argued by analogy. By TH. 2, $\underline{\xi}_3$ is aligned with a principal axis along which $\underline{h}$ has no component. For definiteness, call it axis $\underline{X}_3$. In this case, $\nu = I_3$ in Eq. 9c, and 9a, b take the scalar form

$$(I_\alpha - \lambda)\,\xi_{1\alpha} + J_1\,\xi_{2\alpha} = 0 \qquad (10a)$$

$$J_1\,\xi_{1\alpha} + (I_\alpha - \mu)\,\xi_{2\alpha} = 0 \qquad (10b)$$

(not summed on α). Eliminating $\xi_{2\alpha}$ necessary and sufficient equilibrium conditions are

$$\left[(I_\alpha - \lambda)(I_\alpha - \mu) - J_1^2\right]\xi_{1\alpha} = 0 \qquad (11)$$

(not summed). Now $\xi_{13} = 0$ because $\xi_{33} = 1$, and neither ξ_{11} nor ξ_{12} are zero, for otherwise $\underline{\xi}_1$ is a principal axis contrary to TH. 2. Thus for $\alpha = 1, 2$ the bracketed quantity in Eq. 11 must vanish.

Suppose $D = I_1 - I_2 \neq 0$. Then

$$\lambda = \frac{1}{2}\left(I_1 + I_2 \pm \sqrt{D^2 - 4J_1^2}\right) \qquad (12a)$$

$$\mu = \frac{1}{2}\left(I_1 + I_2 \mp \sqrt{D^2 - 4J_1^2}\right) \qquad (12b)$$

Evidently: (*)

TH. 4 If $|I_1 - I_2| < 2\,|J_1|$, no real equilibrium exists for which $\underline{h}$ is in the plane of the local horizontal (i.e. the $\xi_1 \xi_2$ plane). Here I_1 , I_2 are principal plane containing $\underline{h}$.

Refer to Fig. 1 for the definition of angles Θ, φ and put $H = |\underline{h}| / \omega_0 = \sqrt{J_\alpha J_\alpha}$. The angle φ and magnitude H are considered known, and J_1 , Θ are to be determined from Eqs. 8d, 10a (with $\alpha = 1$, say): i.e. from $J_1 = H \cos(\Theta + \pi/2 - \varphi)$ and $\frac{1}{2} \quad D \mp \overline{D^2 - 4J_1^2} \cos\Theta + J_1 \sin\Theta = 0$. Define

(13a)
$$\varkappa = -D/2H = (I_2 - I_1)/2H$$

By labeling axes so $I_2 \geq I_1$ we also can assure $\varkappa \geq 0$. Then it is straightforward to show that Θ is to be determined from

(14)
$$\sin(\Theta + \varphi) = \varkappa \sin 2\Theta$$

This establishes the qualitative relationship between φ and Θ shown in Fig. 2. If $0 \leq \varkappa \leq \frac{1}{2}$ $(0 \leq I_2 - I_1 \leq H)$ there is ex-

---

(*) This has an immediate analogue for the case where $J_1 = 0, J_3 \neq 0$ again calling $\underline{X}_3$ the axis with no $\underline{h}$-component:

TH.5 If $(I_1 - I_2| < 2|J_3/4)$, no real equilibrium exists which $\underline{h}$ is in the plane normal to the orbit (i.e. the $\xi_2 \xi_3$ plane).

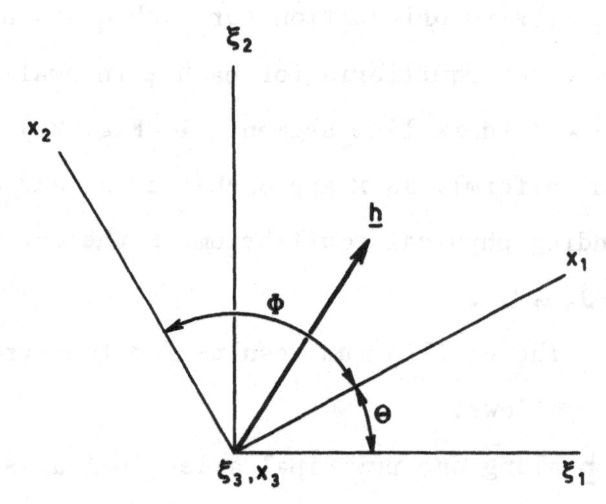

Fig.1. Rotor Angular Momentum in the Plane of Local Horizontal

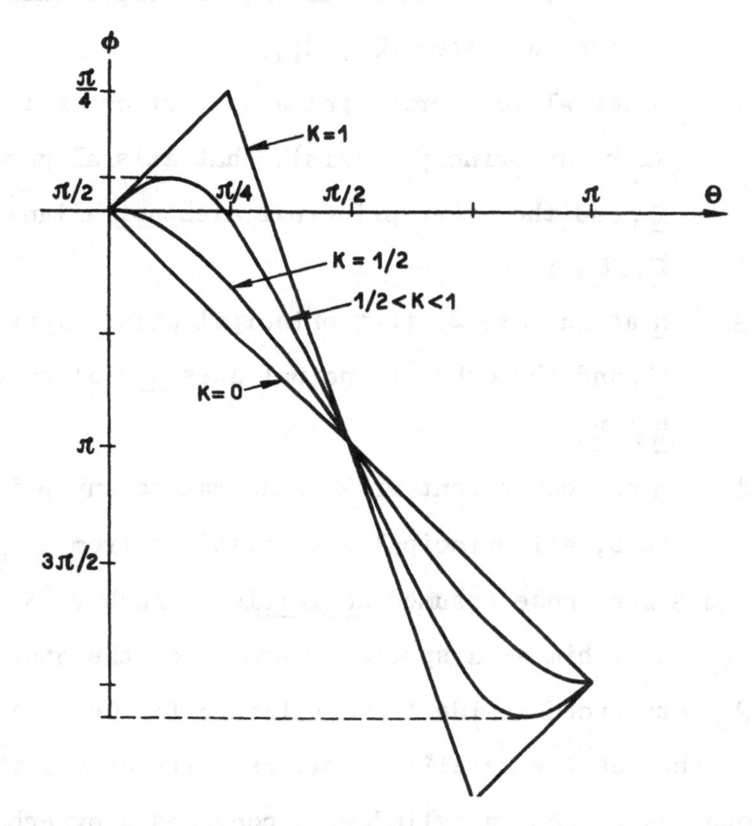

Fig. 2. Solution Possibilities for Equation 13

actly one equilibrium orientation for each φ-value. For $1/2 < \varkappa \leq 1$ there are two equilibria for each φ in small ranges near $\varphi = 0$ and $\varphi = -\pi$ (dark line segments in Fig. 2.) The case $D = 0$ is approached uniformly as $\varkappa$ approaches zero, and one finds that the corresponding physical equilibrium is the one found previously with $J_1 = J_3 = 0$.

The equilibrium results for the gyrostat can be summarized as follows.

Case 1. $\underline{h}$ along one principal axis, that axis in turn align with $\pm\underline{\xi}_2$ and the remaining principal axes aligned in any way with $\pm\underline{\xi}_1, \pm\underline{\xi}_3$;

Case 2. $\underline{h}$ normal to a principal axis (but not coincident with any principal axis), that axis aligned with $\underline{\xi}_1$ and the other principal axes <u>not</u> aligned with $\underline{\xi}_2, \underline{\xi}_3$;

Case 3. $\underline{h}$ as in Case 2, that principal axis aligned with $\underline{\xi}_3$ and the other principal axes <u>not</u> aligned with $\underline{\xi}_1, \underline{\xi}_2$;

Case 4. $\underline{h}$ not coincident with or normal to any principal axis, all principal axes distinct from $\underline{\xi}_1, \underline{\xi}_2, \underline{\xi}_3$.

Cases 2 and 3 are those assumed <u>ab initio</u> by Anchev [5] with Case 1 treated by him as a special instance of the others. Likins [3] introduced an ideal nomenclature for Cases 1, 2 and 3. Observe that as the satellite traverses its orbit, the axis of the rotor sweeps out a cylinder, a cone and a hyperbola of one sheet in these three cases respectively. This is shown in Fig. 3, adapted from [4]. We shall use this nomenclature interchangeably with the case number hereafter: Case 1, cylindrical case; Case 2, conical case; Case 3 hyperbolic case. Figures 4 and 5 are summary depictions of the rotor in relation to ξ -frame for the four basic cases.

It is shown in [6] that Eq. 14 is the basis for establishing the equilibrium orientation in both Cases 2 and 3, wether considering the restricted problem as at present or the unrestricted problem. Only the value of $\varkappa$ is different. It is given for Case 2 by Eq. 13a, and

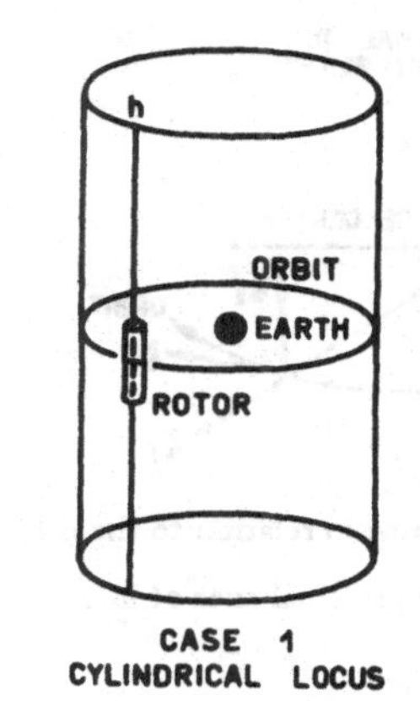

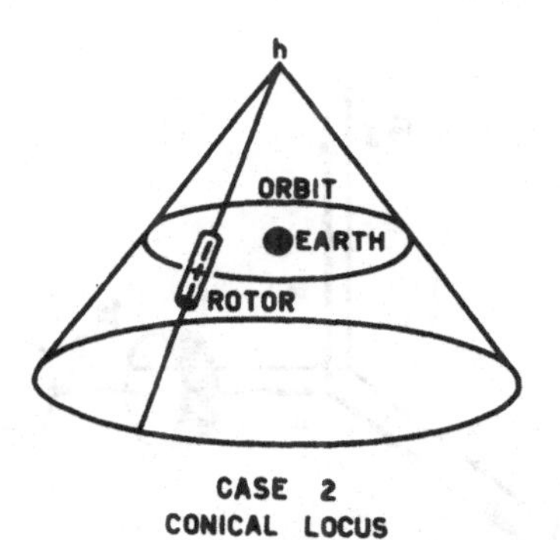

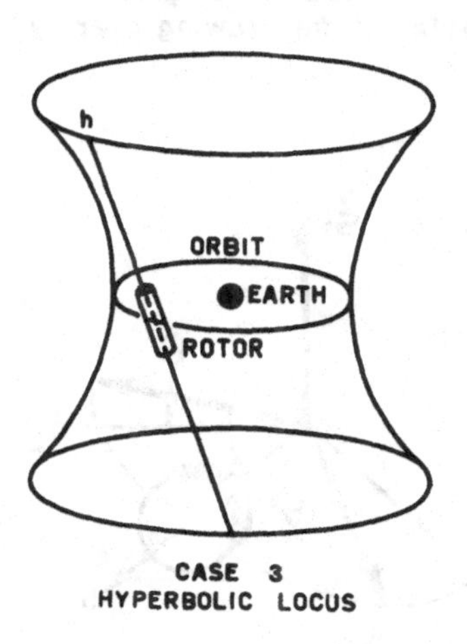

Fig. 3. Geometric properties of Spin-axis for case 1, 2, 3 equilibria

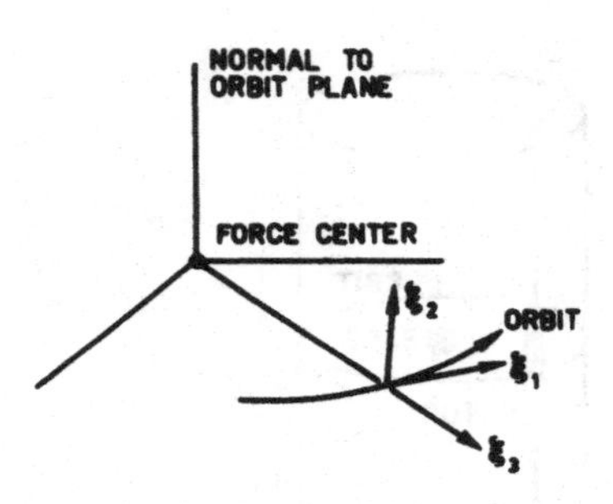

Fig. 4. The ξ-frame in relation to the orbit
Heavy frames are principal axes of inertia, labeling arbitrary

for Case 2 by

$$\varkappa = 2\,(I_3 - I_2)/H \tag{13b}$$

Let us consider the solution of Eq. 14 in somewhat more detail than is given in Fig. 2.

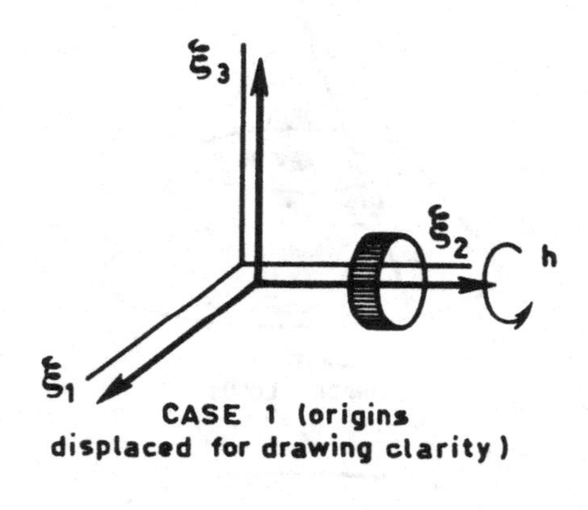

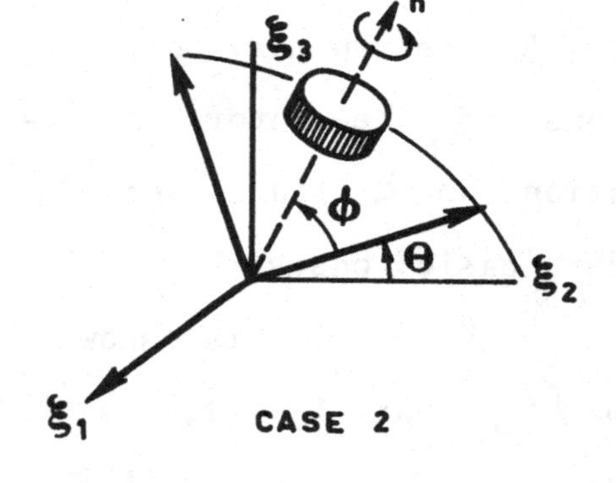

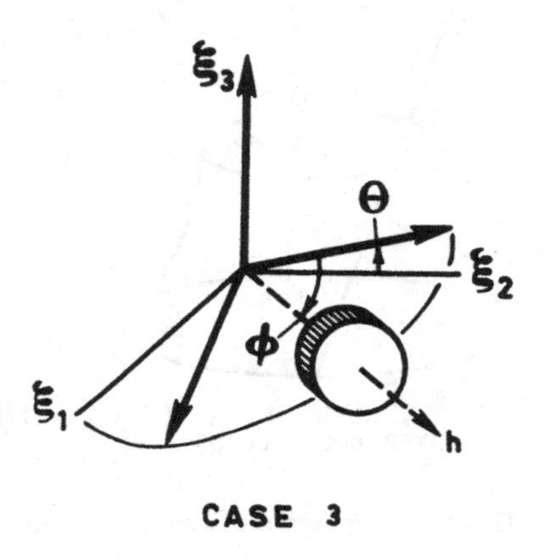

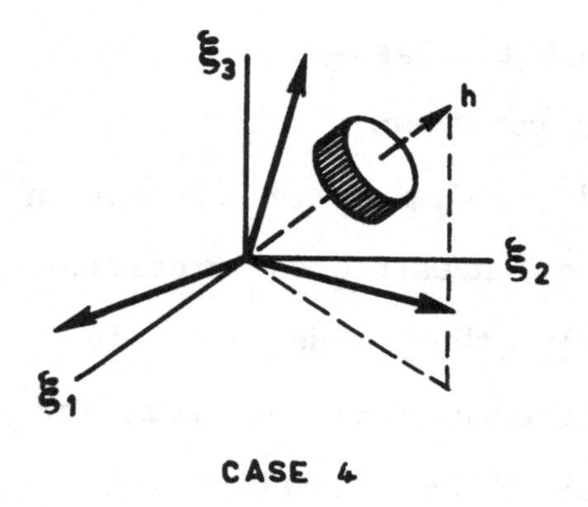

Fig. 5. The basic equilibrium cases

Following 6 ,however, we remove the restriction following Eq. 13a that $\varkappa$ be positive, and now assume the body axes can be labeled arbitrarily and that $\varkappa$ can range freely over $(-\infty,\infty)$. This parameter, which establishes the character of the solutions, measures the difference in principal moments on the axes normal to the rotation axis, in relation to the angular momentum of the internal rotor (actually an inertia-like parameter equal to that angular momentum divided by the orbit angular frequency). As a first observation we note that the periodicity of the trigonometric functions divides the $\varphi - \theta$ plane into small modules of a few basic types. Figure 6 shows such a division. The modules labeled I and II, contained within the heavy border, can be considered basic. Within the modules labeled I and II the same transcendental equation is satisfied, except that the algebraic sign of $\varkappa$ is reversed. Thus, if curves of φ vs. θ for typical $\varkappa$ in the range $(-\infty,\infty)$ are obtained in modules I and II, the same curves apply in I' and II' but are labeled with $\varkappa$ -values having the opposite sign from those used in the original modules. The region $|\theta| \leq \pi/2$, $0 \leq \varphi \leq \pi$ is shown in Fig. 7 for several $\varkappa$ -values, to show how the various curves fit together. As a practical matter, one can costruct the figure by solving Eq. 14 for φ as a function of θ , obviating the need to treat it as transcendental in the dependent variable. Actually, though, it is more normal to be given φ and to wish to derive θ , the latter to be used as inputs to further relationships connected with

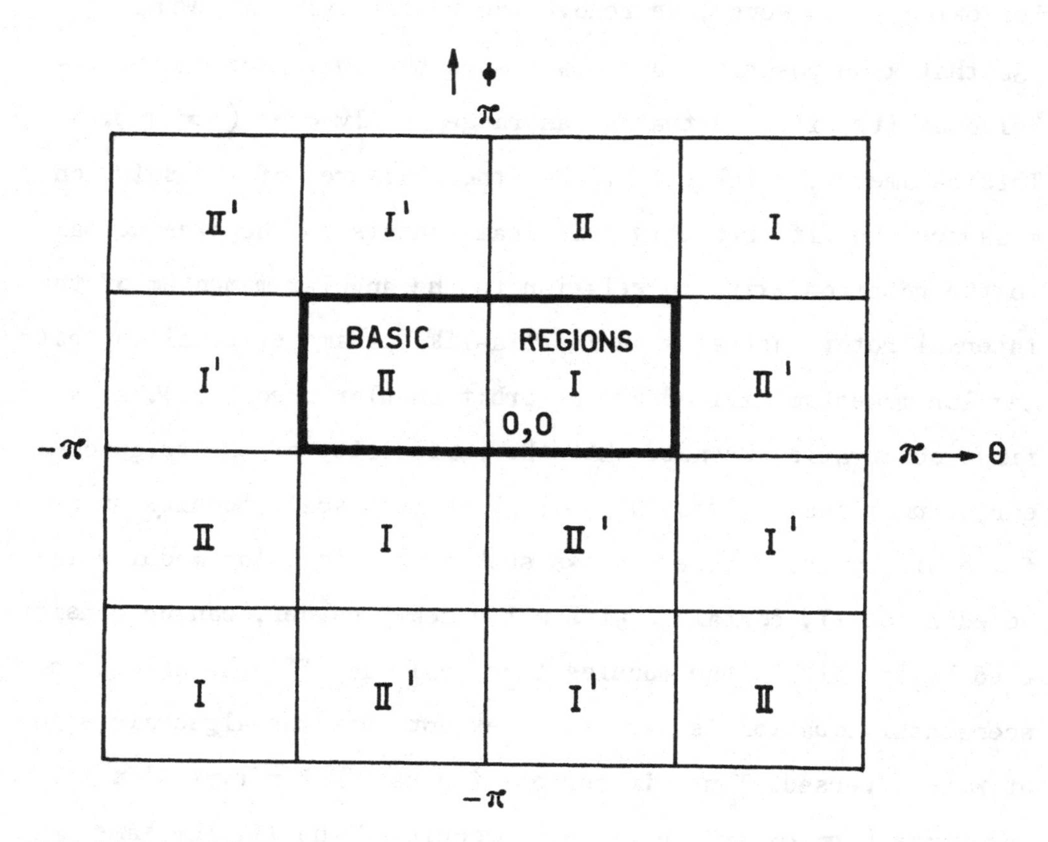

Fig. 6. Solution Modules for Eq. 14

the stability properties of the gyrostats.

It is straightforward to construct a numerical (computer) program for the solution of Eq. 14, the only ramifications being a need to keep straight the inverse trigonometric functions and the number of solutions (one, two or three) as they depend on $\varkappa$.

The following characteristics of the solutions should be noted. For $0 \leq \varkappa \leq 1/2$, there is only one Θ-value

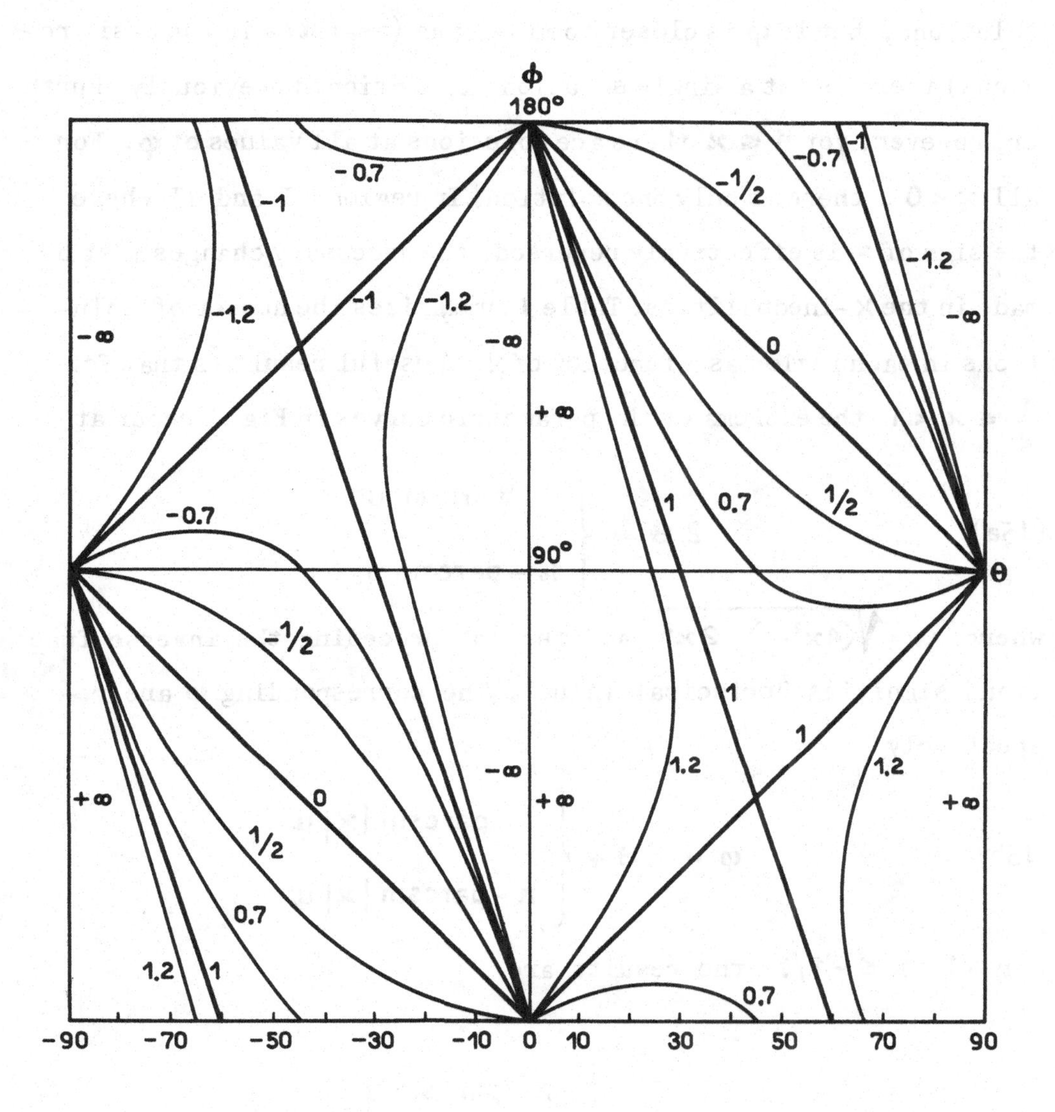

Fig. 7. Solutions of Eq. 14 for constant ϰ.

satisfying Eq. 14 for each preassigned φ-value. For $\frac{1}{2} < \varkappa < 1$, if φ is close enough to Θ or $\pi/2$ (in the basic regions) there are three solutions, but if φ is closer to mid-range (near $\pi/4$ in the basic regions) there is but a single solution, as mentioned previously. Further, however, for $1 \leq \varkappa$ there are solutions at all values of φ. For all $\varkappa < 0$, there is only one solution. In regions I and II where the sign of $\varkappa$ is effectively reversed, the necessary changes must be made in the $\varkappa$-inequalities. Table 1 summarizes the number of solutions in each region as a function of $\varkappa$. A useful result is that for $\frac{1}{2} \leq \varkappa < 1$ the extrema of the parametric curves in Fig. 7 occur at

$$(15a) \qquad 2\Theta = \begin{cases} \text{parcsin } u \\ \pi - \text{parcsin } u \end{cases}$$

where $u = \sqrt{(4\varkappa^2-1) \quad 2\varkappa^2}$ and the "p" preceding the inverse functions signifies "principal value". The corresponding φ are respectively

$$(15b) \qquad \varphi = -\Theta + \begin{cases} \text{parcsin } |\varkappa| \, u \\ \pi - \text{parcsin } |\varkappa| \, u \end{cases}$$

For $-1 < \varkappa \leq -1/2$ the results are

$$(16) \qquad 2\Theta = \begin{cases} -\pi + \text{parcsin } u \\ -\text{parcsin } u \end{cases}$$

Table 1 Number of Solutions of Eq. 14

	Region			
	I	II	I'	II'
$\varkappa \leq -1$	0	1	2	1
$-1 < \varkappa < -1/2$	0	1	2 (note a) 0 (note b)	1
$-1/2 \leq \varkappa < 0$	0	1	0	1
$0 < \varkappa \leq 1/2$	0	1	0	1
$1/2 < \varkappa \leq 1$	2 (note a) 0 (note b)	1	0	1
$1 < \varkappa$	2	1	0	1

Notes

[a] φ near extremes of module

[b] φ near mid-range of module

with φ again given by Eq. 15b.

In conclusion, we see that for either Case 2 or Case 3, the nature of the orientation equilibria in the restricted problem is completely determined by a single parameter $\varkappa$ which represents the ratio of the difference of two principal moments of inertia to an "effective inertia parameter" H of the internal rotor (specifically, the rotor angular momentum with respect to the body divided by the orbit angular velocity). Depending on the value of $\varkappa$ (in the range $-\infty, \infty$) there can be one equilibrium or three equilibria. Case 1 can be subsumed within Case 2 by putting the angle φ between the rotor axis and the gyrostat X_2 body axis equal to zero. The multiple solutions for equilibria corresponding to prescribed φ and $\varkappa$ have very important implications to the engineering applications of orbiting gyrostats.

References

[1] Thomson, W.T., "Spin stabilization of attitude against gravity torque," J. Astronaut. Sci. 9 (1962), 31-33.

[2] Kane, T.R., Marsh, E.L. and Wilson, W.G., Letter to the editor, J. Astronaut. Sci. 9 (1962), 108-109.

[3] Likins, P.W., "Stability of a symmetrical satellite in attitudes fixed in an orbiting reference frame," J. Astronaut. Sci. 12 (1965), 18-24.

[4] Kane, T.R. and Mingori, D.L., "Effect of a rotor on the attitude stability of a satellite in a circular orbit," AIAA J. 3 (1965), 936-940.

[5] Anchev, A.A., "Flywheel stabilization of relative equilibrium of a satellite, "Kosmicheskie Issledovaniya 4 (1966), 192-202.

[6] Roberson, R.E., "Equilibria of Orbiting Gyrostats," J. Astronaut. Sci. 15 (1968), 242-248.

[7] Likins, P.W. and Roberson, R.E., "Uniqueness of Equilibrium Attitudes for Earth-Pointing Satellites," J. Astronaut. Sci. 13 (1966), 87-88.

[8] Roberson, R.E. and Hooker W.W., "Gravitational equilibria of a rigid body containing symmetric rotors," Proc. 17th Congr. Int. Astronaut. Fed. (Madrid, 1966) Dunod, Paris, 1967.

9. Stability of the Special Equilibria

Introduction

We consider now the stability of Cases 2 and 3 discussed in the previous lecture, with that Case 1 falling out as a by-product. The essential symbolic preliminaries are recapitulated first.

Case 2 ("conical case")

(1a) $$h/\omega_0 = H\left[0 \quad \cos\varphi \quad \sin\varphi\right]^T$$

(1b) $$\underline{X}_\alpha = {}^1A_{\alpha\beta}(\Theta)\underline{\xi}_\beta$$

(1c) $$\varkappa = 2(I_3 - I_2)/H$$

Case 3 ("hyperbolic case")

(2a) $$h/\omega_0 = H\left[-\sin\varphi \quad \cos\varphi \quad 0\right]^T$$

(2b) $$\underline{X}_\alpha = {}^3A_{\alpha\beta}(\Theta)\underline{\xi}_\beta$$

(2c) $$\varkappa = (I_1 - I_2)/2H$$

Then in each case an equilibrium angle Θ is given by a solution of the transcendental equation

(3) $$\sin(\Theta + \varphi) = \varkappa \sin 2\Theta$$

Case 1 ("cylindrical case") is a limiting form of both Case 2 and Case 3 in which $\varphi = 0$ and $\varkappa = \pm\infty$.

At least three tools are available to apply to the determination of the stability of any particular equilibrium orientation. The first of these, historically speaking, is what is usually referred to as Chetaev's method. (Actually, in [1] Chetaev gives an example and illustrates the technique, but leaves it to the reader to infer the rationale of its application to more general situations.) In this method one starts with known integrals of the motion, say $V_1, \ldots, V_n$, and forms linear (*) combination $V = \Sigma a_i V_i + C$ where C is an arbitrary constant. It is clear that V is constant during the motion; moreover C can be so chosen so that $V = 0$ when the system is in the equilibrium state. If the coordinates of the perturbed motion now are inserted in V and the quadratic approximation in small deviations from the equilibrium is retained, the fundamental theorem on stability of Liapunov assures that the positive definiteness of the quadratic form is a sufficient condition for the stability of the equilibrium under consideration. The success of this method depends, first, on the explicit availability of the integrals; and second, on the user's skill in manipulating the combination V_i into a pure quadratic form. This route is taken by Anchev [2]. He finds sufficient conditions for stability in Cases 1 and 3,

---

(*) In practice it may be necessary to add quadratic terms, but since $\frac{d}{dt}(V_i V_j) = 0$ if $\frac{dV_i}{dt}$ and $\frac{dV_j}{dt}$ are, the argument which follows is not changed.

but for Case 2 only in a restricted perturbation. In the application of this method later in the present paper his sufficient result for Case 3 is weakened and Case 2 is treated for a general perturbation.

The second method is based on the recognition of the Hamiltonian function as an integral for certain dynamical systems, and its decomposition into a positive definite form in the generalized velocities plus a so-called "dynamic potential energy" function. Establishing the conditions under which the latter is positive provides sufficient conditions for stability. If cyclic coordinates are present in the problem one must first remove them and construct the Hamiltonian from the Routhian rather than the Lagrangian function. This second route has been taken by Rumiantsev [3], [4]. He obtains a dynamic potential energy function $W = -R_0$, where R_0 is the portion of the Routhian function independent of the generalized coordinate derivatives. (in the absence of cyclic coordinates, the dynamic potential energy function is the negative of the corresponding portion of the Lagrangian.) He first obtains equilibrium conditions by forming $W_1 = W + \ell F$, where ℓ is a Lagrange multiplier and F is a constraint relationship among the more-than-minimal set of generalized coordinates he uses to describe W, and by equating to zero the partial derivatives of W_1 with respect to those coordinates. He then examines the second partial derivatives and finds sets of inequalities that are supposed to assure W_1 positive definite (i.e. minimum dynamic potential energy) at the several equilibria. The conditions he applies on the second derivatives to assure stability, however, appear to be applied without further reference to the fact that the four variables he uses are not independent. This means, in effect, that he is specifying more stringent conditions on the second partials than are really necessary when the constraint is recognized. It follows that he obtains sufficient conditions for stability, but not as weak sufficient conditions as are inherently available in the method.

The third approach is via the Thomson-Tait-Chetaev theorem. Suppose that Θ is a 3 by 1 matrix which represents the small angular deviation of the body axes from their equilibrium orientation in the reference frame. By a technique described in [5] a linear dynamical equation is easily developed in the form.

$$I\ddot{\Theta} + \omega_0 G\dot{\Theta} + \omega_0^2 K\Theta = 0 \tag{4}$$

where G is a skew-symmetric and K is a symmetric matrix, the structure of these two matrices depending on the specific equilibrium being examined. The rationale of the method is as follows. A theorem originally due to Thomson and Tait (proved by Chetaev, more recently resurrected by Zajac, and embellished by Zajac and Pringle) (*) states that if the equation above also

(*) See [6] for a review of pertinent literature.

contains a damping term $D\dot{\Theta}$ with D symmetric and positive definite, (*) the null solution is stable or unstable according as K is positive definite or sign variable respectively. In other words, the matrix G plays no role in establishing the stability of a "sufficiently damped" system: one need examine only K. To be sure, we have no damping here, but the result suggests that the stable equilibria found from a consideration of K alone will be the only stable equilibria that are physically meaningful, inasmuch as "real" systems inevitably include some dissipative mechanisms. This approach, therefore, consists in finding an expression for K in the several equilibrium cases and determining the conditiions on the parameters that it be positive definite.

It should be noted that all three viewpoints lead only to sufficient conditions for stability: the first because it involves the construction of a Liapunov function and an application of the Liapunov stability theorem which furnishes only sufficient conditions; the second for the same reason, with the Hamiltonian function directly providing the Liapunov function; the third, because it does not take account of possible stabilization by gyroscopic terms. However, both necessary and sufficient conditions for infinitesimal stability (not necessarily Liapunov stability) can be found from Eq. 4 if one finds the

---

(*) It sometimes is possible to relax this to semi-definiteness.

eigenvalues of the matrix A obtained when it is rewritten in the first order canonical form $\dot{y} = Ay$. These eigenvalues, of course, depend on G as well as K .

We begin with the matrix dynamical equation

$$I\dot{\omega} + \tilde{\omega}(I\omega + h) = 3\omega_0^2 \tilde{\xi}_3 I \xi_3 \tag{5a}$$

and the kinematical differential equations

$$\dot{\xi}_\alpha = -\tilde{\Omega}\xi_\alpha \quad (\alpha = 1,2,3) \tag{5b}$$

The torque term on the right hand side of Eq. 5a can be derived from a potential function

$$U = \frac{mk}{2r^3} \operatorname{tr} I - \frac{3}{2}\omega_0^2 \xi_3^T I \xi_3 \tag{6}$$

but we work here only with the portion $-3\omega_0^2 \xi^T I \xi_3/2$ which depends on rotation.

Consider the functions

$$f = \omega^T I \omega - 2U_1 \qquad g = \xi_2^T (I\omega + h) \tag{7a-b}$$

Differentiating them with respect to time and eliminating $\dot{\omega}$ and $\dot{\xi}_\alpha$ as these arise by means of Eqs. 5, one finds that

$$\dot{f} = 6\omega_0^3 \xi_3^T I \xi_1 \qquad \dot{g} = 3\omega_0^2 \xi_3^T I \xi_1 \tag{8a-b}$$

Thus neither f nor g is an integral of the motion, (*) but the

(*) Unless, of course, the motion is so restricted that either ξ_1 or ξ_3 is always an eigenvector of I .

combination

(9) $$V_1 = f - 2\omega_0 g = \omega^T I\omega - 2U_1 - 2\omega_0 \xi_2^T (I\omega + h)$$

is such an integral. (*)

Other integrals are kinematical, reflecting the orthonormal relationship of the ξ_α :

(10a-b-c) $$V_2 = \xi_2^T \xi_2 \qquad V_3 = \xi_3^T \xi_3 \qquad V_4 = \xi_2^T \xi_3$$

all are integrals of the motion. (They have the constant values 1, 1 and 0 respectively.)

In addition to the equilibrium condition given as Eq. 3, we shall find it convenient to refer to the more general form derived as Eqs. 9 of the previous lecture, namely

(11a) $$(I - \lambda E)\xi_1 + j_1 \xi_2 = 0$$

(11b) $$j_1 \xi_1 + (I - \mu E)\xi_2 + \frac{1}{4} j_3 \xi_3 = 0$$

(11c) $$\frac{1}{4} j_3 \xi_2 + (I - \varrho E)\xi_3 = 0$$

where (**), recall, λ, μ, ϱ are eigenvalue-like constants and j_α are the projections of $\underline{h}/\omega_0$ on $\underline{\xi}_\alpha$, i.e., $\underline{h}/\omega_0 = j_\alpha \underline{\xi}_\alpha$.

(*) Anchev gives these functions separately as integrals, but in his application he always uses them in a combination equivalent to V_1 .

(**) Minor modifications in the notation are adopted: ϱ and j_α replace the ν and J_α of [8] to avoid confusion with other uses of the latter symbols in [1].

Application of Chetaev's Method

Denote with asterisks the equilibrium values of ξ_α and let ϵ , δ_α be matrices of small deviations of ω and ξ_α respectively from equilibrium. Form the combination of integrals (*)

$$V = V_1 + \omega_0^2 (\mu + j_2) V_2 - 3\omega_0^2 \varrho V_3 + \frac{3}{2}\omega_0^2 j_3 V_4 + C \tag{12}$$

It is understood that the constants j_2, j_3, μ, ϱ are those pertaining to equilibrium. Use $\xi_\alpha = \xi_\alpha^* + \delta_\alpha$ and $\omega = \omega_0 \xi_2^* + \epsilon$ in this expression and choose C to remove all terms independent of δ_α and ϵ . Retain all terms in the expansion. In this way one obtains

$$V = \epsilon^T I \epsilon - 2\omega_0 \delta_2^T I \epsilon + 3\omega_0^2 \delta_3^T I \delta_3 + \frac{3}{2}\omega_0^2 j_3 \delta_2^T \delta_3 + \\ + \omega_0^2 (\mu + j_2) \delta_2^T \delta_2 - 3\omega_0^2 \varrho \delta_3^T \delta_3 \ . \tag{13}$$

In the cancelation of certain terms during the intermediate steps one makes use of the general equilibrium relations written as Eqs. 11, by using them to replace such forms as $I\xi_2^*$ and $I\xi_3^*$ by linear combinations of the ξ_α^* .

(*) Minor modifications in the notation are adopted: ϱ and j_α replace the ν and J_α of [8] to avoid confusion with other uses of the latter symbols in [1] .

Note especially that the matrix level has not been abandoned in obtaining the last form for V , a fact that facilitates the several groupings performed in the sequel. Let us add to and substract from the right hand side of Eq. 13 the quantity $\omega_0^2 \delta_2^T I \delta_2$. Then V evidently becomes

$$V = (\epsilon - \omega_0 \delta_2)^T I (\epsilon - \omega_0 \delta_2) + \omega_0^2 \left[\delta_2^T (\mu E + j_2 E - I) \delta_2 + \right.$$

$$\left. + \frac{3}{2} j_3 \delta_2^T \delta_3 + 3 \delta_3^T (I - \varrho E) \delta_3 \right] . \tag{14}$$

The first term of which is obviously positive definite. The other can be rewritten as a qudratic form by defining

$$P = (\mu + j_2) E - I \qquad Q = 3 (I - \varrho E) \tag{15a}$$

$$\delta = \begin{bmatrix} \delta_2 \\ \delta_3 \end{bmatrix} \qquad M = \begin{bmatrix} P & \frac{3}{4} j_3 E \\ \frac{3}{4} j_3 E & Q \end{bmatrix} \tag{15b}$$

Thus

$$V = (\epsilon - \omega_0 \delta_2)^T I (\epsilon - \omega_0 \delta_2) + \omega_0^2 \delta^T M \delta \tag{16}$$

A sufficient condition for stability is that the 6 by 6 matrix M be positive definite. We first take a naive approach and suppose that the elements of the 6 by 1 matrix δ are independent. It is a detail to work out M for Case 3: it is a diagonal matrix with

$$M_{11} = j_2 + (I_2 - I_1) \cos^2 \Theta \tag{17a}$$

$$M_{22} = j_2 + (I_1 - I_2)\sin^2\Theta \tag{17b}$$

$$M_{33} = j_2 + I_1\sin^2\Theta + I_2\cos^2\Theta - I_3 \tag{17c}$$

$$M_{44} = 3\,(I_1 - I_3) \tag{17d}$$

$$M_{55} = 3\,(I_2 - I_3) \tag{17e}$$

$$M_{66} = 0 \tag{17f}$$

Superficially it appears troublesome that $M_{66} = 0$, but it is shown below to cause no problem when the correct approach is used. (We can, in fact, avoid even the suggestion of a problem by modifying the function V by the addition of a term $V_3^2/4$, in which case the term $\xi_3^*\xi_3^{*T}$ is added to Q and $M_{66} = 1$ results. This is unnecessary, it turns out.) Equations 17 (excepting 17f) lead to sufficient conditions for stability

$$M_{11} > 0\,,\quad M_{22} > 0,\quad M_{44} > 0,\quad M_{55} > 0 \tag{18}$$

inasmuch as the first two inequalities also imply $M_{33} > 0$. Using Eqs. 2c and 3 to eliminate $I_1 - I_2$ from Eqs. 17a and 17b, the latter also can be written as stability conditions, $-j\sin\varphi/\sin\Theta > 0$ and $j\cos\varphi/\cos\Theta > 0$ respectively. The remaining conditions agree directly with Anchev.

Clearly these are valid sufficient conditions, but Anchev goes on to show for Case 1 (using a different Liapunov function) that sufficient conditions for this case are $I_2 - I_1 + j > 0$ and $I_2 - I_3 + j_2 > 0$, together with $M_{44} > 0$, $M_{55} > 0$.

Of these the second is implied by the others, whence Case 1 conditions reduce to

$$j_2 > I_1 - I_2 \,, \quad I_1 > I_3 \,, \quad I_2 > I_3 \tag{19}$$

On the other hand, a direct specialization of Eqs. 18 to Case 1 gives Eq. 19 together with $j_2 > 0$. Since the latter inequality apparently is unnecessarily stringent, one feels it should be possible to weaken Eqs. 18 as sufficient requirements for Case 3 itself. The key, of course, is the fact that the elements of matrix δ are not independent, so theform $\delta^T M \delta$ can be positive without insisting that the entire matrix M be positive definite. Recall that $V_2 = 1$, $V_3 = 1$ and $V_4 = 0$. These together imply $\xi^*_{2\alpha}\delta_{2\alpha} = 0$, $\xi^*_{3\alpha}\delta_{3\alpha} = 0$, $\xi^*_{2\alpha}\delta_{2\alpha} = 0$ to within second order terms (which become of fourth order in the quadratic form). In particular:

Case 2

$$\delta_{22} = \delta_{32} \tan\Theta, \quad \delta_{32} = -\delta_{23}\,, \quad \delta_{33} = \delta_{23} \tan\Theta \tag{20}$$

whence

$$\begin{aligned} \delta^T M \delta = {} & P_{11}(\delta_{21})^2 + \frac{3}{2} j_3 \delta_{21}\delta_{31} + Q_{11}(\delta_{31})^2 + \\ & + \left(P_{22}\tan^2\Theta + P_{33} + Q_{22} + Q_{33}\tan^2\Theta\right)(\delta_{23})^2 \end{aligned} \tag{21}$$

to a quadratic approximation. Sufficient conditions for stability are

$$P_{11} + Q_{11} > 0 \tag{22a}$$

$$P_{11} Q_{11} > (3j_3/4)^2 \tag{22b}$$

$$(P_{22} + Q_{33}) \tan^2\Theta + (P_{33} + Q_{22}) > 0 \tag{22c}$$

where, for this case,

$$P_{11} = j_2 + \mu - I_1 = j_2 + I_2 \cos^2 + I_3 \sin^2\Theta - I_1 \tag{23a}$$

$$Q_{11} = I_1 - \varrho = I_1 - I_2 \sin^2\Theta - I_3 \cos^2\Theta \tag{23b}$$

$$P_{22} = j_2 + \mu - I_2 = j_2 + (I_3 - I_2)\sin^2\Theta \tag{23c}$$

$$P_{33} = j_2 + \mu - I_3 = j_2 + (I_2 - I_3)\cos^2\Theta \tag{23d}$$

$$Q_{22} = 3(I_2 - I_3)\cos^2\Theta \tag{23e}$$

$$Q_{33} = 3(I_3 - I_2)\sin^2\Theta \tag{23f}$$

Using these in Eq. 22c, the latter takes the simpler form

$$j_2 + 4(I_2 - I_3)\cos 2\Theta > 0 \tag{24a}$$

it can be seen that Eqs. 24ab are both implied by the single inequality

$$P_{11} + Q_{11} > \sqrt{(P_{11} - Q_{11})^2 + (3j_3/2)^2} \tag{24b}$$

Equations 24 are sufficient for Case 2 stability. For the moment, no attempt is made to further simplify Eq. 24b.

Case 3

$$\delta_{22} = -\tan\Theta\,\delta_{21}\,,\quad \delta_{33} = 0\,,\quad \delta_{23} + \delta_{31}\sin\Theta + \delta_{32}\cos\Theta = 0$$

whence, also recalling $j_3 = 0$.

$$\delta^T M \delta = (P_{11} + P_{22}\tan^2\Theta)(\delta_{21})^2 +$$

$$+ (Q_{11} + P_{33}\sin^2\Theta)(\delta_{31})^2 + 2P_{33}\sin\Theta\cos\Theta\,\delta_{31}\delta_{32} +$$

$$+ (Q_{22} + P_{33}\cos^2\Theta)(\delta_{32})^2 \tag{26}$$

to a quadratic approximation. Sufficient conditions for stability are

$$P_{11} + P_{22}\tan^2\Theta > 0 \tag{27a}$$

$$Q_{11} + Q_{22} + P_{33} > 0 \tag{27b}$$

$$Q_{11}Q_{22} + P_{33}(Q_{11}\cos^2\Theta + Q_{22}\sin^2\Theta) > 0 \tag{27c}$$

where for Case 3

$$P_{11} = \mu + j_2 - I_1 = j_2 + (I_2 - I_1)\cos^2\Theta \tag{28a}$$

$$P_{22} = \mu + j_2 - I_2 = j_2 + (I_1 - I_2)\sin^2\Theta \tag{28b}$$

$$P_{33} = \mu + j_2 - I_3 = j_2 + I_1\sin^2\Theta + I_2\cos^2\Theta - I_3 \tag{28c}$$

$$Q_{11} = 3(I_1 - I_3) \tag{28d}$$

$$Q_{22} = 3(I_2 - I_3)$$

Using Eqs. 28 as needed, Eq. 29a becomes

$$j_2 + (I_2 - I_1)\cos^2\Theta > 0 \tag{29a}$$

Furthermore, the single inequality

$$Q_{11} + Q_{22} + P_{33} > \sqrt{(Q_{11}-Q_{22})^2 + 2\,(Q_{22}-Q_{11})P_{33}\cos 2\theta + P_{33}^2} \tag{29b}$$

implies both Eqs. 27bc. Equations 29 are sufficient for Case 3 stability.

Application of Dynamic Potential Energy Method

The dynamic potential energy of the system in rotation is

$$W = U_1 - T_0 \tag{30}$$

where U_1 is ordinary potential energy. (In accordance with astronomical tradition, Eq. 6 gives a potential fucntion which is the negative of potential energy.) Also, T_0 is the part of the kinetic energy that is independent of the derivatives of generalized coordinates. Inasmuch as

$$T = \frac{1}{2}\,\omega^T I\,\omega + \omega^T h + \frac{\dot{\delta}^2}{2 I_r} \tag{31}$$

where I_r is the polar moment of inertia of the internal rotor, and inasmuch as $\omega = \Omega + \omega_0 \xi_2$ with only Ω containing the generalized coordinate derivatives, we can use Eq. 6 and immediately write

(32) $$W = \frac{3}{2}\omega_0^2 \xi_3^T I \xi_3 - \frac{1}{2}\omega_0^2 \xi_2^T I \xi_2 - \omega_0 \xi_2^T h + C$$

Here C is an arbitrary additive constant.

We wish to establish the conditions under which W has a minimum at the equilibria. Suppose that a virtual change in orientation occurs about equilibrium, described by the 3 by 1 matrix $\delta\Theta$ whose elements are independent small angles. Denote by ξ_α^* the values of ξ_α at equilibrium. Then after the virtual rotation one has

$$\left[\xi_1 \;\; \xi_2 \;\; \xi_3\right] = \left[E - \delta\tilde{\Theta} + \frac{1}{2}\delta\tilde{\Theta}\,\delta\tilde{\Theta} + \frac{1}{2}\delta\bar{\Theta}\, U_0\, \delta\bar{\Theta} + \dots\right]\left[\xi_1^* \;\; \xi_2^* \;\; \xi_3^*\right]$$

(33)

The quadratic approximation to the virtual rotation direction cosine matrix is obtained from [7], in which the definition of constant matrix U_0 is given (*) explicitly. From this follows

(34) $$\xi_\alpha = \xi_\alpha^* - \widetilde{\delta\Theta}\, \xi_\alpha^* + \frac{1}{2}\left(\widetilde{\delta\Theta}\,\widetilde{\delta\Theta} + \overline{\delta\Theta}\, U_0\, \overline{\delta\Theta}\right)\xi_\alpha^*$$

(*) The matrix denoted there by U is here denoted U_0 to avoid confusion with the potential function. The overbar is standardized operational symbol like the tilde in this formalism. It denotes a diagonal matrix whose principal diagonal elements of the column matrix under the overbar.

or

$$\delta\xi_\alpha = \tilde{\xi}^*_\alpha \delta\Theta \tag{35a}$$

$$\delta^2\xi_\alpha = (-\widetilde{\delta\Theta}\,\tilde{\xi}^*_\alpha + \overline{\delta\Theta}\, U_0\, \bar{\xi}^*_\alpha)\delta\Theta \tag{35b}$$

Conditions for an equilibrium are that $\delta W = 0$ there: conditions that it be a minimum are that $\delta^2 W < 0$. We therefore proceed by calculating the first and second variations using Eq. 32 for W, then substituting Eq. 35 in the result as needed. This procedure is carried out entirely on the matrix level. We find

$$\delta W = 3\omega_0^2 \xi_3^T I \delta\xi_3 - \omega_0^2 \xi_2^T I \delta\xi_2 - \omega_0 h^T \delta\xi_2$$

Evaluating this at equilibrium,

$$\delta W = \omega_0^2 \left[3\xi_3^{*T} I \tilde{\xi}_3^* - \xi_2^{*T} I \tilde{\xi}_2^* - \frac{1}{\omega_0} h^T \tilde{\xi}_2^* \right] \delta\Theta \tag{36}$$

whence the condition of equilibrium must be the vanishing of the quantity in square brackets. Setting its transpose equal to zero,

$$\tilde{\xi}_2^* (I\xi_2^* + h/\omega_0) = 3\tilde{\xi}_3^*\, I\xi_3^* \tag{37}$$

This, of course, is precisely the well-known matrix equilibrium condition already obtained by another route in the previous lecture.

The second variation is

$$\delta^2 W = 3\omega_0^2 \xi_3^T I \delta^2\xi_3 - \omega_0(\omega_0 \xi_2^T I + h^T)\delta^2\xi_2 + 3\omega_0^2 \delta\xi_3^T I \delta\xi_3 - \\ - \omega_0^2 \delta\xi_2^T I \delta\xi_2 \tag{38}$$

Substitution for these variations of ξ_α in terms of $\delta\Theta$

$$\delta^2 W = \omega_0^2\, \delta\Theta^T K' \delta\Theta \tag{39a}$$

where

$$K' = 3\left[(\widetilde{I\xi_3^*})\tilde{\xi}_3^* - \tilde{\xi}_3^* I \tilde{\xi}_3^* + \overline{I\xi_3^*}\, U_0 \bar{\xi}_3^*\right] + \left[\xi_2^* I \tilde{\xi}_2^* - (\widetilde{I\xi_2^*} + h/\omega_0)\tilde{\xi}_2^* - \right.$$
$$\left. - (\overline{I\xi_2^*} + \bar{h}/\omega_0) U_0 \bar{\xi}_2^*\right] \tag{40}$$

Using the equilibrium conditions given as Eqs. 37, this can be rearranged as

$$K' = K - (\mu + j_2)\bar{\xi}_2^* U_0 \bar{\xi}^* + 3\varrho\, \bar{\xi}_3^* U_0 \bar{\xi}_3^* -$$
$$- \frac{3}{4} j_3 (\bar{\xi}_2^* U_0 \bar{\xi}_3^* + \bar{\xi}_3^* U_0 \bar{\xi}_2^*) \tag{41}$$

where

$$K = \tilde{\xi}_2^* \left[I - (\mu + j_2)E\right] \tilde{\xi}_2^* + \tilde{\xi}_3^* \left[3(\varrho E - I)\right] \tilde{\xi}_3^* -$$
$$- \frac{3}{4} j_3 (\tilde{\xi}_2^* \tilde{\xi}_3^* + \tilde{\xi}_3^* \tilde{\xi}_2^*) \tag{42}$$

Note that K is a symmetric matrix, whereas the remaining terms in K' are skew-symmetric by virtue of the skew-symmetry of U_0. (See [7] .) This is true regardless of the order in which the virtual angle rotations are performed.) Only K survives in the quadratic form for $\delta^2 W$, so the latter becomes

$$\delta^2 W = \omega_0^2\, \delta\Theta^T K \delta\Theta \tag{39b}$$

Sufficient conditions for stability are that K be positive definite. The matrix K obtained by this route is precisely the matrix to which one is led in following the route of the Thomson-Tait-Chetaev.

Application of Thomson-Tait-Chetaev Theorem

Let the body be in its equilibrium orientation characterized by ξ_α^*, and let it subsequently undergo a small rotation described by the 3 by 1 matrix Θ. We use the matrix linearization procedure described fully in [5] to obtain

$$\omega = \dot{\Theta} + \omega_0 \tilde{\xi}_2^* \Theta + \omega_0 \xi_2^* \tag{43a}$$

$$\xi_3 = \xi_3^* + \tilde{\xi}_3^* \Theta \tag{43b}$$

Substituting these into Eq. 5a, linearizing whenever products of Θ and of its derivatives arise, we very easily (*) and directly obtain Eq. 4 with the explicit form for the K-matrix

$$K = \left(\tilde{\xi}_2^* I - I\tilde{\xi}^* - h/\omega_0\right)\tilde{\xi}_2^* + 3\left(I\xi_3^* - \xi_3^* I\right)\tilde{\xi}_3^* \tag{44}$$

By the rational described following Eq. 4, the matrix G in Eq. 4 is of no consequence to the stability of the "corresponding damped system". Stability is established by the positive definiteness of K alone. Before proceeding, note that the elimination

(*) The case should be carefully noted, and the train of manipulation should be compared with that required to implement the dynamic potential energy method.

of the combinations $I\xi_2^*$ and $I\xi_3^*$ by means of Eq. 11bc leads directly to the expression for the matrix K that arose previously in connection with the dynamic potential method, thereby establishing the identity of these two matrices. It is slightly easier to work with K in the previous form, because it preserves symmetry in an explicit way.

Case 1

$$K = \begin{bmatrix} 4(I_2 - I_3) + j & 0 & 0 \\ 0 & 3(I_1 - I_3) & 0 \\ 0 & 0 & I_2 - I_1 + j_2 \end{bmatrix} \tag{45a}$$

Stability conditions are

$$I_1 > I_3 \qquad j_2 + 4(I_2 - I_3) > 0 \qquad j_2 + (I_2 - I_1) > 0 \tag{46abc}$$

Note that if $j_2 = 0$ (no rotor), these reduce to the stability condition $I_2 > I_1 > I_3$ that holds for a single rigid body within the so-called "Lagrange region" of the various inertia parameter phase planes that are used to represent stability behavior as a function of shape. Because the conditions are only sufficient ones, there is no flaw in the argument because it fails to identify the "Beletski-Delp region" of these planes. The inertia parameter phase plane that one chooses to display the results obviously is not unique, and several actually have been used in the literature. The one used here is a plane of two parameters

$\Gamma_1 = (I_3 - I_2)/I_1$ and $\Gamma_3 = (I_2 - I_1)/I_3$. (Their negatives often are used.) Both of these parameters have physical limitation to the range $[-1, 1]$. Figure 2 shows two examples, the unshaded regions are those permitted by the inequalities in the captions. More detailed examination of the stability regions in this plane is the subject of the next section.

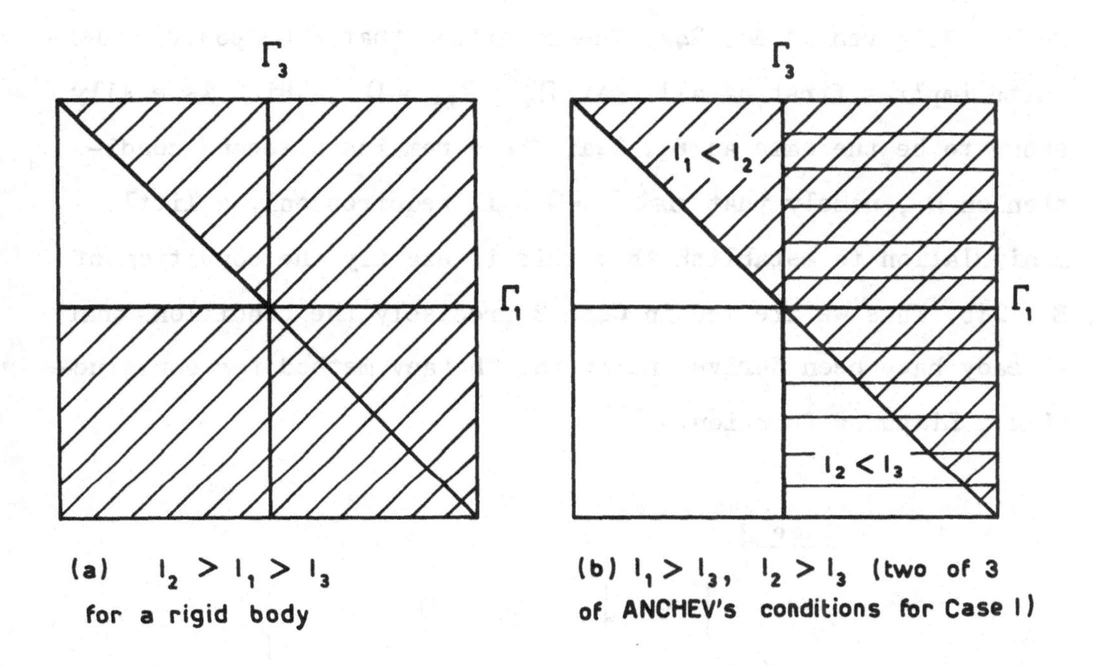

Fig. 1. Inertia Parameter Representation of Stability Results

<u>Case 2</u>

$$K = \left[\begin{array}{c|c} j_2 + 4\,(I_2 - I_3)\cos 2\Theta & 0 \\ \hline 0 & \\ 0 & R \end{array}\right]$$

where R is the 2 by 2 sub-matrix whose elements are

(48a) $$R_{11} = P_{11}\sin^2\Theta + Q_{11}\cos^2\Theta - \frac{3}{4}j_3\sin 2\Theta$$

(48b) $$R_{12} = R_{21} = (P_{11} - Q_{11})\sin\Theta\cos\Theta - \frac{3}{4}j_3\cos 2\Theta$$

(48c) $$R_{22} = P_{11}\cos^2\Theta + Q_{11}\sin^2\Theta + \frac{3}{4}j_3\sin 2\Theta$$

(The P_{11}, Q_{11} are the Case 2-values given by Eq. 23.) The requirement K_{11} be positive is the same condition already obtained by Method 1, given as Eq. 24a. The condition that R be positive definite implies first of all that $R_{11} + R_{22} > 0$, which is easily shown to be the same as Eq. 24a. There remains a second condition on R, namely that $\det R > 0$. It requires only a little manipulation to establish that this is exactly the condition of Eq. 24b. Thus we are led in Case 2 precisely the conditions that already have been derived using the Chetaev method for constructing a Liapunov function.

Case 3

(49) $$K = \left[\begin{array}{c|c} S & \begin{matrix}0\\0\end{matrix} \\ \hline 0 & j_2 + (I_2 - I_1)\cos 2\Theta \end{array}\right]$$

where S is the 2 by 2 sub-matrix whose elements are

(50a) $$S_{11} = Q_{22} + P_{33}\cos^2\Theta$$

(50b) $$S_{12} = S_{21} = -P_{33}\sin\Theta\cos\Theta$$

$$S_{22} = Q_{11} + P_{33} \sin^2 \Theta \tag{50c}$$

In which P_{33} , Q_{22}, Q_{11} are the Case 3 values given by Eqs. 28. It is easily verified that sufficient conditions for the positive-definiteness of the K -matrix are again precisely the same as already obtained by the first method of this work, specifically by Eqs. 27 or their equivalents Eqs. 29.

Discussion

Sufficient conditions now have been obtained by three routes: by Chetaev's method for constructing a Liapunov function; by the method of minimizing the synamic potential; by applying the Thompson-Tait-Chetaev Theorem. The second method reduces to an examination of the positive definiteness of the same matrix as arises in the third method, but the approach using the Thompson-Tait-Chetaev theorem is far less laborous than the others to carry through in this particular problem.

Figure 3 shows a stability region for Case 1 in a $\Gamma_1 \Gamma_3$ -plane, using a dimensionless ratio $\eta = j / I_2$ was used for "rotor strength," for a typical positive value of this ratio, $\eta = 0{,}3$. It is seen that the same curve $j_2 = I_1 - I_2$ is a bounding condition in both cases but that it extends into the fourth quadrant in the present method, whereas the condition

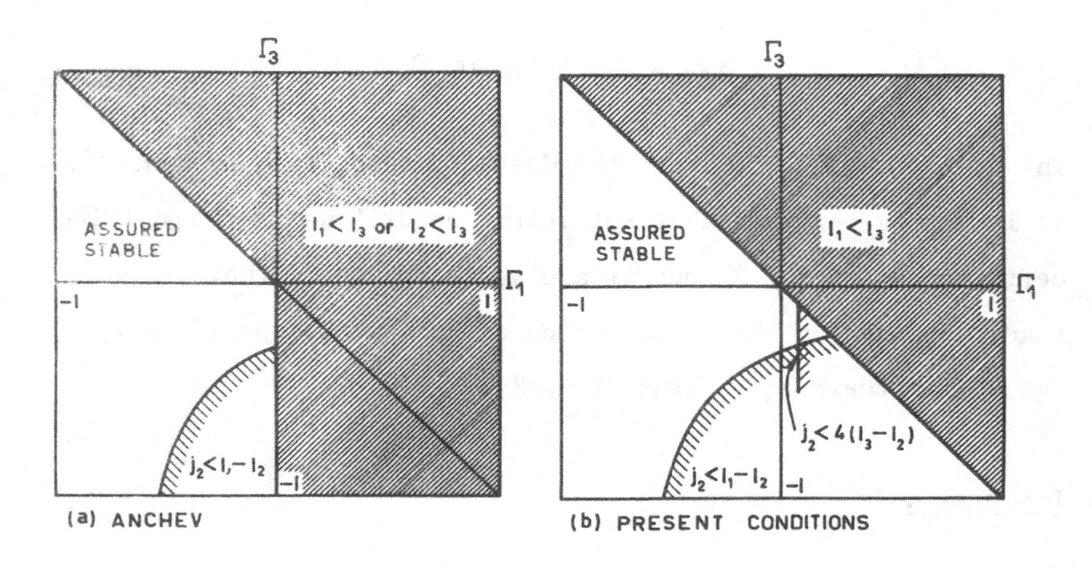

Fig. 2. Comparison of Sufficient Conditions for Case 1 (j/I = 0.3)

$I_2 > I_3$ prevents any fourth quadrant area to be assured stable using Anchev's criteria. The existence of the small piece of assured-stable area in the present case is important because this region is a part of the gyroscopically stabilized Beletski-Delp region for the orbiting rigid body. One would infer, therefore, that although a rigid body in a configuration appropriate to the Beletski-Delp region $(I_1 > I_2 > I_3)$ cannot be stable in the presence of damping, body of the same shape sometimes can be made stable by the addition of a rotor aligned with the axis of minimum moment of inertia such that it would remain stable in the presence of dissipation. This could have significant engineering implications in some instances.

Figure 4 illustrates the change in Case 1 stability region with increasing η. Only the third and fourth quadrants of the $\Gamma_1 \Gamma_3$-plane are shown, corresponding to $\eta > 0$. The figure also includes the boundary of the Beletski-Delp region for reference. Corresponding results for negative η are illustrated in Fig. 5. As would be expected, the stability region draws in with more negative η, until $\eta = -0{,}5$ there is none left. Physically, this just means that the body tends to flip around into an orientation for which $\eta = +0{,}5$ is stable.

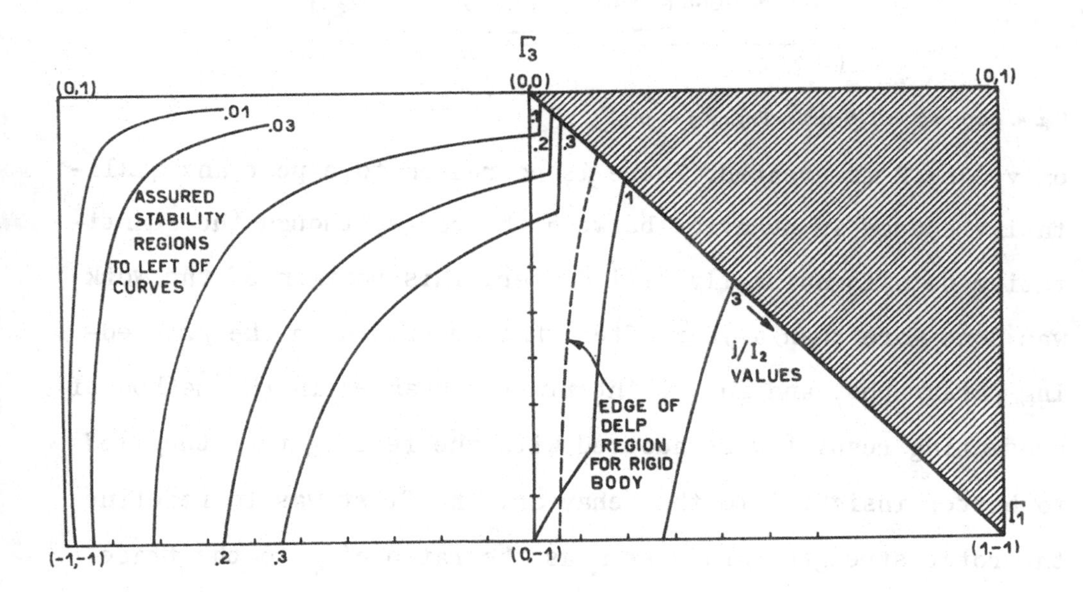

Fig. 3. Edge of Case 1. Stability Region, $j > 0$

Having described the behavior of the stability region for Case 1 ($\varphi = 0$) and changing rotor strength, it remains to illustrate the cases for which $\varphi \neq 0$. In what follows

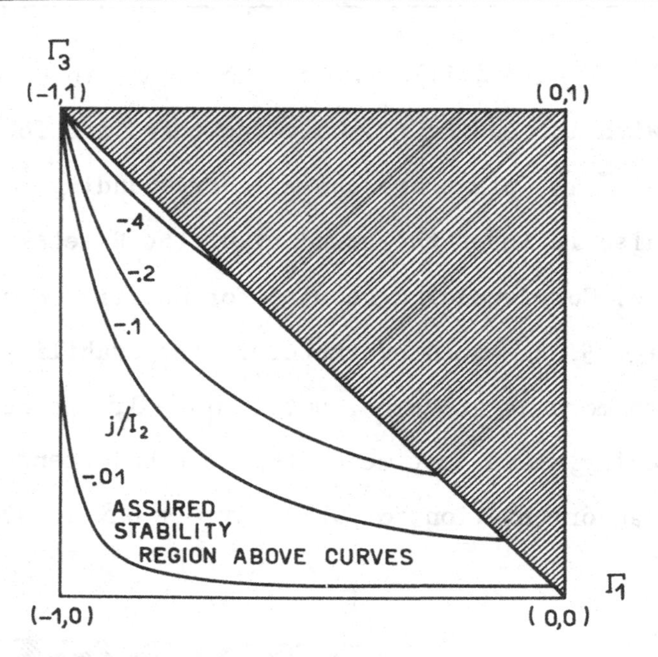

Fig. 4. Edge of Case 1. Stability Region, $j < 0$

only Case 2 is treated. There is no reason to expect any qualitative change in behavior between the cases, though the quantitative details naturally will differ. This portion of the work was done more than a year after that described in the preceeding paragraphs, and in the interim two changes in the method of presenting results were adopted with the feeling that they led to better insight into the behavior. The first was to redefine the rotor strength parameter η as the ratio of j to the trace of the inertia matrix of the body. This normalization is a measure of the "size" of the body, and it seems more sensible to hold size fixed while varying shape over the inertia phase plane than to vary both at once. The second change, even more significant, is the abandonment of the $\Gamma_1\,\Gamma_3$-plane in favor of Magnus' triangular inertia phase plane. The virtues of the latter are manifold, once any initial distaste for a triangular phase region is overcome. Because this representation is not yet universally known. Fig. 6 is included to make clear how the points of the triangle are related to the principal moments of inertia.

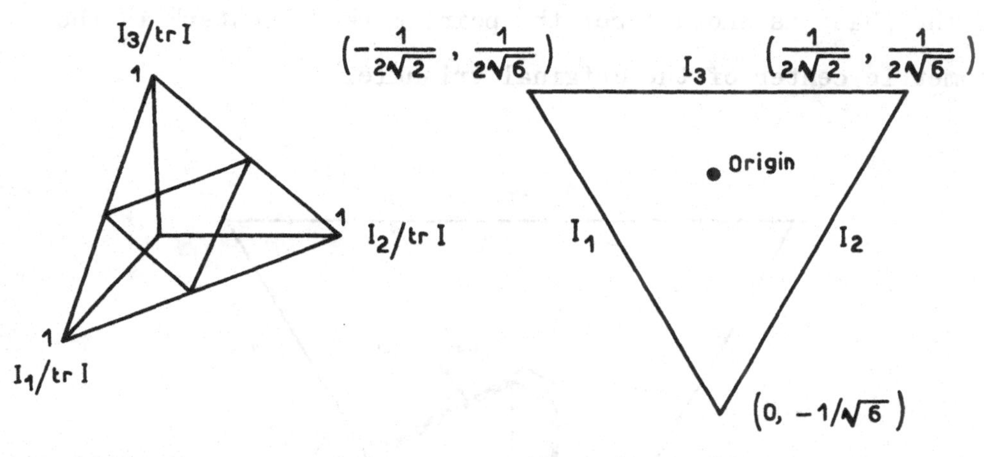

Fig. 5. The Magnus Inertia Phase Triangle

I have chosen to normalize to trace I, so the slanted plane in Fig. 6a has the equation $I_1 + I_2 + I_3 = 1$. Shaded regions (e. g. $(I_3 > I_1 + I_2)$ are not physically realizable. Figure 6b is the final inertia phase triangle, showing rectangular cartesian coordinates of the vertices referred to an origin at the center of the equilateral triangle. The values of the ratios $I_\alpha / \mathrm{tr}\, I$ to coordinates measured from this point are easy to work out.

To give a checkpoint with what has gone before, we can reproduce the case shown in Fig. 3, but now using the inertia triangle. The result is Fig. 7. Using this method of representation the behavior of the stable region in Case 2 is shown in Fig. 8 as a function of angle φ for several values of η. To magnify the important part of the picture only the lower point of the phase is shown here: the point marked "center" is the geometric center of the original triangle.

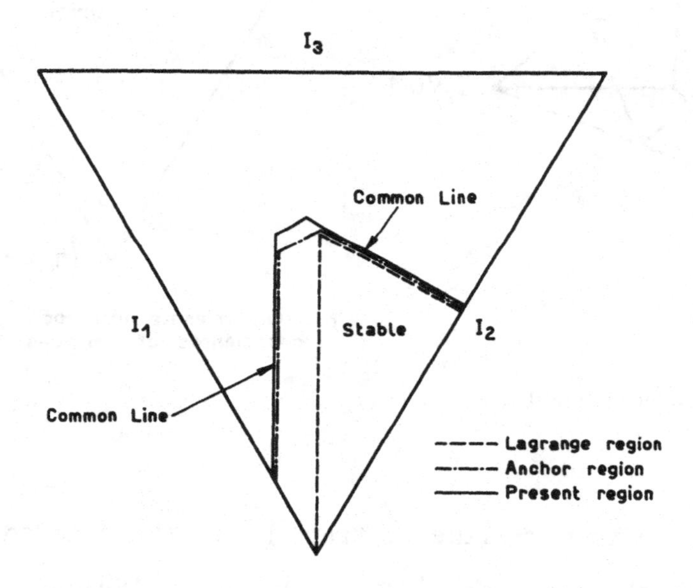

Fig. 6. Case 1. Stability Region, Analogous to Figure 3, in Inertia Phase Plane

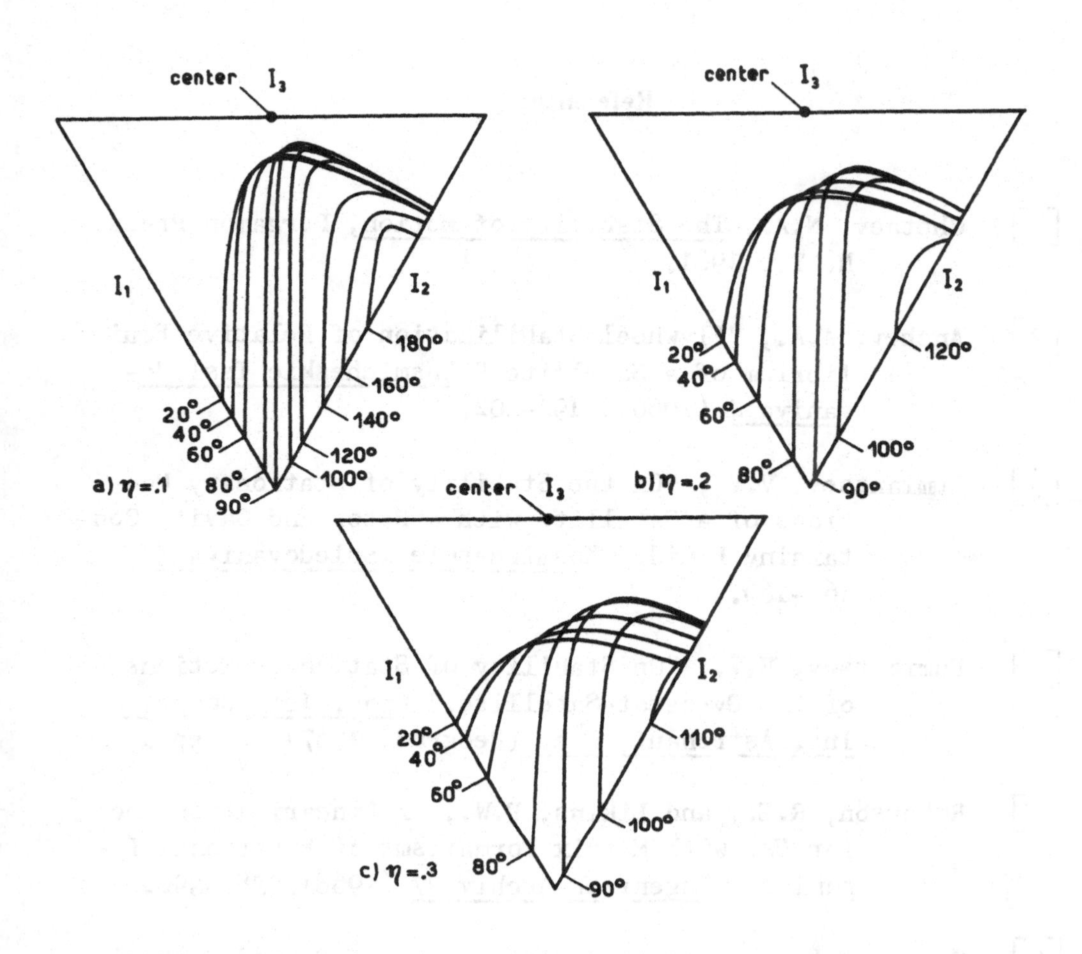

Fig. 7. Case 2 Stability Regions

References

[1] Chetaev, N.A., The Stability of Motion, Pergamon Press, N. Y., 1961.

[2] Anchev, A.A., "Flywheel Stabilization of Relative Equilibrium of a Satellite," Kosmicheskie Issledovaniya 4 (1966), 192-202.

[3] Rumiantsev, V.V., "On the Stability of Stationary Motions of a Satellite with a Rotor and Cavity Containing Fluid," Kosmicheskie Issledovaniya 5 163-169.

[4] Rumiantsev, V.V., "On Stability of Stationary Motions of the Gyrostat-Satellite," Proc. 18th Congr. Int. Astronaut. Fed. (Beograd, 1967) to appear.

[5] Roberson, R.E., and Likins, P.W., "A Linearization Tool for Use with Matrix Formalisms of Rotational Dynamics," Ingenieur-Archiv 37 (1968), 388-392.

[6] Zajac, E.E., "Comments on 'Stability of Damped Mechanical Systems' and a Further Extension," AIAA J. 3 (1965), 1749-1750.

[7] Likins, P.W. and Roberson, R.E., "The Quadratic Approximation in Rotational Dynamical Equations," Ingenieur-Archiv 38 (1969), 53-57.

10. General Equilibria of Gravitationally Stabilized Gyrostats

In this lecture we consider eqyilibria for Case 4. where neither $J_1 = 0$ nor $J_3 = 0$. The problem of finding these equilibria can be approached in two different ways:

1. One can look for all orentations of a given body which there exists a relative angular momentum vector that makes the orientation an equilibrium.

2. Or, one can assume as given a body with a specified relative angular momentum vector and look for all possible equilibrium orientations.

The former is especially appealing in the synthesis problem where one desires to implement a specified equilibrium orientation. The latter is the more natural viewpoint if the satellite construction is already specified and it is desired only to determine what equilibrium it might attain. In the previous lectures the second viewpoint was implicitly adopted, but the first viewpoint is the more fruitful one in arriving at the general solution. In theory, knowledge of the general solution for a given body completely specified the solution to the second formulation, but the conversion form one to the other is not a simple matter. The nature of the problem is illustrated in the special cases by the fact that a specified body orientation leads to only two rotor orientations within the body, for a given magnitude of the relative angular momentum, whereas a specified rotor orientation within the body can lead to as many as twenty-four body equilibrium orientations. This lack of one-to-one correspondence also complicated the general problem.

The equilibrium conditions are those exhibited previously,

$$
\begin{aligned}
(I-\lambda E)\xi_1 &= -J_1\xi_2 \\
(I-\mu E)\xi_2 &= -J_1\xi_1 - \frac{1}{4}J_3\xi_3 \\
(I-\nu E)\xi_3 &= -\frac{1}{4}J_3\xi_2
\end{aligned}
\tag{1}
$$

Inner products of Eqs. (1) with ξ_1, ξ_2, and ξ_3 give the following equivalent set of equations:

$$
\begin{aligned}
\xi_1^T I \xi_1 &= \lambda & \qquad \xi_2^T I \xi_1 &= -J_1 \\
\xi_2^T I \xi_2 &= \mu & \qquad \xi_3^T I \xi_1 &= 0 \\
\xi_3^T I \xi_3 &= \nu & \qquad \xi_3^T I \xi_2 &= -\frac{1}{4}J_3
\end{aligned}
\tag{2}
$$

The ξ_α must also satisfy the orthonormality conditions, of course. We assume now that $J_1 \neq 0$ and $J_3 \neq 0$.

We first show that:

1. No Case 4 solution exists for an axially symmetric body;
2. No Case 4 solution exists for a principal axis in the $\xi_1\xi_2$-plane;
3. No Case 4 solution exists for a principal axis in the $\xi_3\xi_2$-plane.

If the body is spherically symmetric, all directions are principal axes and the left sides of Eqs. 1 vanish. But the right side does not vanish in Case 4. Suppose next that $I_1 = I_2$. Then $\xi_3^T I \xi_1 = 0$ (Eq. 2) implies that either $\xi_{13} = 0$ or $\xi_{33} = 0$. But if the former, $\xi_2^T I \xi_1 = -J_1$, implies $J_1 = 0$, contrary to hypothesis, whilst if the latter, $\xi_3^T I \xi_2 = -J_{3/4}$ implies $J_3 = 0$, again contrary to hypothesis. A similar argument can be used for $I_1 = I_3$ and $I_2 = I_3$, so the first property is proved. Hereafter, therefore, we can assume that I_α are all different.

If a principal axis is in the $\xi_1\xi_2$ -plane, the ξ_3 vector must be normal to that principal axis and have exactly one zero component. (It cannot have more, for then it would be a principal axis and we would have Case 3.) By the first and third of Eq.1, a zero occurs in the same component of $\underline{\xi}_1$ and $\underline{\xi}_2$, which would imply that the set $\{\underline{\xi}_\alpha\}$ no longer span three-space. This contradicts the fact that these vectors are orthogonal, so the second property is established. The third follows analogously.

Now we consider an equilibrium type Longman calls Case 4a, namely that in which a principal axis is in the $\xi_1\xi_3$-plane. If this happens, ξ_2 is normal to that principal axis and contains exactly one zero element, say the αth . By Eq. 1a, either $\lambda = I_\alpha$ or ξ_1 has a zero in its αth element. The latter leads to the same contradiction as previously, so we conclude $\lambda = I_\alpha$ and, from Eq. 1c, $\nu = I_\alpha$ as well. From Eq. 1b it immediately follows that

$$\left[(I-\lambda E)(I-\mu E)(I-\nu E)-J_1^2(I-\nu E)-(J_3^2/16)(I-\lambda E)\right]\xi_2=0 \tag{3}$$

The βth row of Eq. 3 $(\beta \neq \alpha)$ is

$$(I_\beta-I_\alpha)\left[(I_\beta-I_\alpha)(I_\beta-\mu)-J^2\right]\xi_{2\beta}=0 \tag{4}$$

where

$$J^2=J_1^2+(J_3/4)^2 \tag{5}$$

Equation 4 implies

$$\mu=I_\beta+I_\gamma-I_\alpha \qquad (\beta,\gamma \neq \alpha) \tag{6}$$

$$J^2=(I_\beta-I_\alpha)(I_\alpha-I_\gamma) \tag{7a}$$

The latter, in turn, implies that I_α is the <u>intermediate</u> moment of inertia, so we have specifically,

$$J=\sqrt{(I_1-I_2)(I_2-I_3)} \tag{7b}$$

Without loss in generality, we now assume that axes are labeled such that the intermediate moment of inertia is I_2 . Write the equilibrium value of ξ_2 as

$$\xi_2=\left[\cos\Theta \quad 0 \quad \sin\Theta\right]^T \tag{8}$$

and temporarily denote ξ_{32} by x . Putting these into Eqs. 1, we find

$$\xi_1 = \begin{bmatrix} -J_1 \cos\Theta / (I_1 - I_2) \\ -x J_3 / 4 J_1 \\ J_1 \sin\Theta / (I_2 - I_3) \end{bmatrix} \tag{9a}$$

$$\xi_3 = \begin{bmatrix} -(J_3 \;\; 4) \cos\Theta / (I_1 - I_2) \\ x \\ (J_3 \;\; 4) \sin\Theta / (I_2 - I_3) \end{bmatrix} \tag{9b}$$

The requirements $\xi_1^T \xi_2 = 0$ and $\xi_3^T \xi_3 = 1$ imply

$$\sin\Theta = \pm \sqrt{|I_2 - I_3|} / J \tag{10a}$$

$$\cos\Theta = \pm \sqrt{|I_1 - I_2|} / J \tag{10b}$$

$$x = \pm J_1 / J \tag{10c}$$

where any choice of signs is permissible which is consistent with the right-handedness of the ξ -frame. The requirement $\tilde{\xi}_2 \xi_3 = \xi_1$ shows that

$$\operatorname{sgn}(x \tan\Theta) = \operatorname{sgn} J_1 \operatorname{sgn}(I_1 - I_3) \tag{11}$$

We can regard Θ as basic: it can be in any of the four quadrants, but within that quadrant it is unique according to Eq. 10ab. Once fixed, the magnitude of x is fixed by Eq. 10c and its sign by Eq. 11. One then returns to Eqs. 8 and 9 to construct the complete relationship between the ξ-frame and the X -frame.

Suppose that X -frame is given in terms of the

ξ -frame by a direction cosine matrix parametrized in Tait-Bryan angles as

$$X = {}^2A(\theta)\,{}^1A(\varphi)\,{}^3A(\pi/2)\xi \tag{12}$$

It is easy to show by writing out the matrix product

$$A = \begin{bmatrix} -\sin\theta\sin\varphi & \cos\theta & -\sin\theta\cos\varphi \\ -\cos\varphi & 0 & \sin\varphi \\ \cos\theta\sin\varphi & \sin\theta & \cos\theta\cos\varphi \end{bmatrix} \tag{13}$$

that

$$\sin\varphi = x \ , \quad \cos\varphi = xJ_3/4J_1 \tag{14}$$

We can visualize the change in orientation with varying φ very easily. Both X_1 and X_3 describe cones about $\underline{\xi}_2$, whilst X_2 remains in the $\xi_1\xi_3$ -plane.

For each combination of J_1, J_3 (i.e. each angle φ), and each body shape, the equilibrium is determined, as well as the projections of $\underline{h}$ on the $\underline{\xi}_1$ and $\underline{\xi}_3$ axes. Hence the $\underline{h}$ -vector is determined in body axes.

To discuss other general equilibria, which Longman calls Case 4b, we use $X = A\xi$ and rewrite Eqs. 2 as

$$A^T I A = \begin{bmatrix} \lambda & -J_1 & 0 \\ -J_1 & \mu & -J_3/4 \\ 0 & -J_3/4 & \nu \end{bmatrix} \tag{15}$$

We can also write this as $I = ABA^T$, and can make the following observations about these equations:

1. I and B are related by an orthogonal similarity transformation. Thus B is an inertia matrix for some set of coordinates. The fact that B is symmetric is no restriction on A . Note that for the synthesis problem the values of J_1, J_3, λ, μ, ν are all arbitrary except that they must be chosen so that B has eigenvalues I_1, I_2, I_3 .

2. The matrix A is the matrix which diagonalizes B . The rows of A are the eigenvectors of the B matrix. Since B is symmetric, the eigenvectors associated with distinct eigenvalues are orthogonal, and if the eigenvalues are not distinct, we are at liberty to choose orthogonal eigenvectors. The rows of a matrix are orthonormal if and only if the columns are. Thus, the requirements that the columns of A be orthonormal is no restriction on Eq. 15.

3. Given any orientation of body axes one can construct the corresponding A matrix. Then it is a simple matter to determine if that orientation can be made into an equilibrium by proper choice of J_1 and J_3 . In particular, we can substitute

a rotation matrix describing a rotation through angle Θ about the ξ_1 axis from the body principal axes to the ξ frame. We obtain immediately that Case 2 orientations exist and require that $J_3 = 4\,(I_\alpha - I_\beta) \times \sin\Theta\cos\Theta$, where α and β depend on which axis coincides with ξ_1 . Similarly, Case 3 is obtained by choosing A as a pure rotation matrix about ξ_3 and we find that $J_1 = (I_\alpha - I_\beta) \times \sin\Theta\cos\Theta$. For Case 1, by property choosing our body axes, we can let $A = E$ and we see immediately that $J_1 = J_3 = 0$.

We can conclude from Eq. 15 that a necessary and sufficient condition that an orientation can be made an equilibrium orientation by proper choice of the relative angular momentum is that the term in the first row third column of B , be zero. In other words

$$\xi_1^T I\, \xi_3 = 0\,. \qquad (16)$$

11. Effects of Flexibility on Stability

We are interested in the attitude of the space vehicle with respect to a rotating reference frame. For a gravity stabilized vehicle this reference will be the orbital frame with one axis perpendicular to the plane of the orbit $\hat{\underset{\sim}{a}}_3$ and a second axis aligned with the local vertical, $\hat{\underset{\sim}{a}}_1$ (Fig. 1). For satellites spinning in free-space, the reference frame will have a constant angular velocity equal to the nominal angular velocity of the system $\underset{\sim}{\omega}_0 = \omega_0 \hat{\underset{\sim}{a}}_3$.

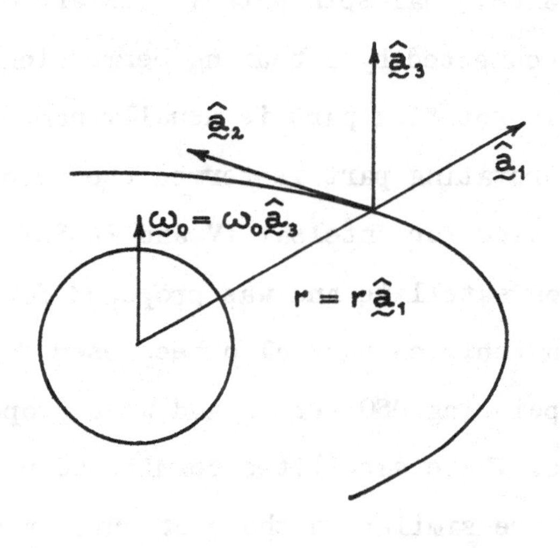

Fig. 1

The description of the behavior of these vehicles form rigid body considerations is generally not sufficient, the effect of energy dissipation being ignored. This energy dissipation can be due to the flexibility of the system in the sense of continuum mechanics or can be obtained from damping mechanisms specially devised to eliminate the nutation. The effect of damping is even more important when the system includes internal angular momenta or when energy is dissipated in parts having different angular rates, as it is the case for multiple-spin vehicles.

These kind of satellites find application in missions for which the ability to point some instruments or a platform and to obtain large gyroscopic stability is required. For instance, dual-spin satellistes are made up of two parts which are connected by a bearing permitting relative rotation. The slowly rotating part is usually named the platform, while the fast rotating part is termed the rotor. Such a configuration has been used for Intelsat IV and TACSAT, the U.S. tactical communication satellite and was proposed for television satellites. Dual--spin vehicles have also been used for scientific missions in the sun-pointing OSO series and were proposed for Ionospheric satellites. These satellites permit the use of techniques and hardware that are similar to those of pure spinners and as it will be seen, permit passive stabilization for satellites with large length to diameter ratios. This has real advantages for the launching operation and could not be achieved with classical satellites.

The rotor is gnerally devised to have an axis of symmetry parallel to the bearing axis. A dual-spin satellite can then be idealized as a rigid gyrostat. A further step is to consider the system as a deformable or flexible gyrostat if it is supposed that deformations take place in one part only.

A deformable gyrostat is defined as a deformable or flexible body containing some rigid constant speed rotors. The variable defining the angular position of the rotors will be considered as cyclic variables and their associated momenta are then supposed constant. It will also be assumed that the rotors are symmetric rotors spinning about thir axes of symmetry. Then the inertia configuration of the system does not depend on the cyclic variables.

Our aim will not be to find all the equilibrium configurations of a given flexible body, but rather to verify if a given body configuration will be compatible with a proposed equilibrium orientation and if this equilibrium is stable.

The system will be said in equilibrium with respect to the reference frame when all the non cyclic internal variables are equal to zero and when some "body-fixed" frame has the same angular velocity as the reference frame.

The time derivative of the total angular momentum of the body with respect to some point is equal to the total torque applied about this point, when the point under consideration is fixed in inertial space or coincides with the centre of mass of the body (this relation is also exact in other very specific circumstances). When the reference point is the centre of mass, we have the vectorial relation.

$$\dot{\underset{\sim}{H}} = \underset{\sim}{L} \,, \tag{1}$$

where $\underset{\sim}{H}$ is the total angular momentum with respect to the centre of mass,

$\underset{\sim}{L}$ is the resultant of external torques applied about the centre of mass.

By definition, $\underset{\sim}{H}$ is the integral over all the elements (dm) of the moment of momentum, or,

$$\underset{\sim}{H} = \int_{mt} \underset{\sim}{\rho} \times \dot{\underset{\sim}{\rho}} \, dm \quad , \tag{2}$$

being the position vector of dm with respect to the centre of mass. The operator $(\dot{\ })$ applied to a vector means time derivative (with respect to some inertial space), mt is the set of elements of mass. The mass is considered as a measure and the formulation remains valid if the system is composed of continuous media, point masses, rigid parts or any of their combinaisons.

The vector will be expressed in the "body-fixed" frame $\{\hat{\underset{\sim}{X}}_\alpha\}$ which will coincide with the reference frame at equilibrium. Then,

$$\underset{\sim}{H} = [\hat{\underset{\sim}{X}}_\alpha]^T H \quad , \quad \underset{\sim}{\rho} = [\hat{\underset{\sim}{X}}_\alpha]^T \rho \quad , \tag{3}$$

where the vector array $[\hat{\underset{\sim}{X}}_\alpha]$ is $[\hat{\underset{\sim}{X}}_\alpha] = [\hat{\underset{\sim}{X}}_1 \ \ \hat{\underset{\sim}{X}}_2 \ \ \hat{\underset{\sim}{X}}_3]^T$. The elements of the matrices H and ρ are then the components of $\underset{\sim}{H}$ and $\underset{\sim}{\rho}$ in the vector basis $\{\hat{\underset{\sim}{X}}_\alpha\}$.

The time derivative of ρ will be

$$\dot{\underset{\sim}{\rho}} = \underset{\sim}{\omega} \times \underset{\sim}{\rho} + \overset{\circ}{\underset{\sim}{\rho}} = [\hat{\underset{\sim}{X}}_\alpha]^T (\tilde{\omega}\rho + \dot{\rho}) \tag{4}$$

where the vector $\overset{\circ}{\underset{\sim}{\rho}}$ is defined as the vector with components equal to the time derivative of the elements of the matrix ρ or

$$\overset{\circ}{\underset{\sim}{\rho}} = [\hat{\underset{\sim}{X}}_\alpha]^T \overset{\circ}{\rho} , \tag{5}$$

this vector is referred to as the "relative velocity" vector of with respect to the frame $\{\hat{\underset{\sim}{X}}_\alpha\}$ and the matrix

$$\tilde{\omega} = \begin{bmatrix} 0 & -\omega_3 & \omega_2 \\ \omega_3 & 0 & -\omega_1 \\ -\omega_2 & \omega_1 & 0 \end{bmatrix} \tag{6}$$

The angular momentum vector can then be written,

$$\underset{\sim}{H} = \int \underset{\sim}{\rho} \times \overset{\circ}{\rho}\, dm + \int \underset{\sim}{\rho} \times \underset{\sim}{\omega} \times \underset{\sim}{\rho}\, dm \ . \tag{7}$$

One defines the inertia tensor of the body as

$$\underset{\sim}{J} = \int (\underset{\sim}{\rho} \cdot \underset{\sim}{\rho}) \underset{\sim}{E} - dm \qquad \underset{\sim}{J} = [\hat{\underset{\sim}{X}}_\alpha]^T J [\hat{\underset{\sim}{X}}_\alpha] , \tag{8}$$

where $\underset{\sim}{E}$ is a unit tensor and J is the inertia matrix of the system (with respect to the centre of mass) expressed in the $\{\hat{\underset{\sim}{X}}_\alpha\}$ basis and the internal angular momentum vector $\underset{\sim}{h}$ is defined as

$$\underset{\sim}{h} = [\hat{\underset{\sim}{X}}_\alpha]^T h = \int \underset{\sim}{\rho} \times \overset{\circ}{\underset{\sim}{\rho}}\, dm \tag{9}$$

The internal angular momentum may include the angular momentum of rotors spinning at constant rates, with respect to the main structure. When the rotors are symmetric, and are spinning with relative angular velocity $\underset{\sim}{\Omega}$ about the axis of symmetry, the corresponding internal angular momentum is equal to

$$\underset{\sim}{h}_r = I'\underset{\sim}{\Omega} = [\hat{\underset{\sim}{X}}_\alpha]^T h_r \tag{10}$$

where I' is the moment of inertia about the axis of symmetry and the norm of $\underset{\sim}{\Omega}$ is a constant. The angular velocity becomes then

$$\underset{\sim}{H} = \underset{\sim}{J} \cdot \underset{\sim}{\omega} + h \quad \text{or} \quad \underset{\sim}{H} = [\hat{\underset{\sim}{X}}_\alpha]^T (J\omega + h) = [\hat{\underset{\sim}{X}}_\alpha]^T H \, . \tag{11}$$

With

$$\underset{\sim}{L} = [\hat{\underset{\sim}{X}}_\alpha]^T L \, , \tag{12}$$

the basic relation (1) becomes,

$$L = \dot{H} + \tilde{\omega} H \quad \text{or} \quad L = J\dot{\omega} + \tilde{\omega} J \omega + \tilde{\omega} h + \dot{J}\omega + \dot{h} \tag{13}$$

This can also be written,

$$\underset{\sim}{L} = \underset{\sim}{J} \cdot \dot{\omega} + \underset{\sim}{\omega} \times \underset{\sim}{J} \cdot \underset{\sim}{\omega} + \underset{\sim}{\omega} \times \underset{\sim}{h} + \overset{\circ}{\underset{\sim}{J}} \cdot \underset{\sim}{\omega} + \overset{\circ}{\underset{\sim}{h}} \, . \tag{14}$$

Here too, the external torques can be considered as a vectorial measure or the torque distribution can be considered as a measurable function of the set of mass elements.

Here also, the gravitational torque in an inverse square field for a satellite on a circular orbit is equal to

$$L = 3\omega_0^2 \hat{\underset{\sim}{a}}_1 \times \underset{\sim}{J} \cdot \hat{\underset{\sim}{a}}_1 \tag{15}$$

where ω_0 is the orbital angular velocity,

$\hat{\underset{\sim}{a}}_1$ is the unit vector aligned with the local vertical.

If θ_{11}, θ_{12} and θ_{13} are the components of $\hat{\underset{\sim}{a}}_1$ in the frame $\{\hat{\underset{\sim}{X}}_\alpha\}$ and if, $\theta_1 = [\theta_{11} \quad \theta_{12} \quad \theta_{13}]^T$, the corresponding matrix L is

$$L = 3\omega_0^2 \tilde{\theta}_1 J \theta_1 \tag{16}$$

At equilibrium, the system is an equivalent rigid gyrostat and if the tensors h_r and $\underset{\sim}{J}$ are expressed in the reference frame $\{\hat{\underset{\sim}{a}}_\alpha\}$ as

$$\underset{\sim}{h}_r = \omega_0 J_\alpha \underset{\sim}{a}_\alpha \quad , \qquad \underset{\sim}{J} = I_{\alpha\beta} \hat{\underset{\sim}{a}}_\alpha \hat{\underset{\sim}{a}}_\beta = [\hat{\underset{\sim}{a}}_\alpha]^T I [\hat{\underset{\sim}{a}}_\alpha]. \tag{17}$$

The following relations have to hold

$$I_{12} = \alpha \quad , \qquad I_{23} = -J_2 \quad , \qquad I_{13} = -\alpha' J_1 \quad , \tag{18}$$

with $\alpha' = 1$ for spin stabilized satellites and $\alpha' = 1/4$ for gravity stabilized vehicles. On orbit α must be equal to zero and in free space α can be set equal to zero without lacking generality. The inertia matrix of the equivalent rigid system is then,

$$I = \begin{bmatrix} I_{11} & 0 & -\alpha' J_1 \\ 0 & I_{22} & -J_2 \\ -\alpha' J_1 & -J_2 & I_{33} \end{bmatrix} \tag{19}$$

We already said that the body-fixed frame will coincide with the reference frame at equilibrium, but we still have to define this frame with respect to the body itself. The reference axes can be the actual principal axes of the system or axes such that the internal momentum vector $\underset{\sim}{h}$ is always equal to zero (Tisserand axes). But the orientation of these axes is defined by rather impractical equations. When deformations can be described by normal modes, one could take as reference the frame which follows the rigid modes, i.e. which is normal to the modes of deformation, but this implies that the deformations can be linearized. Systems composed of interconnected rigid bodies can not be described by this method.

Another "body-fixed" frame, convenient for the determination of the attitude stability of deformable satellites and gyrostats, by the use of linearized equations, is a frame centered at the centre of mass, which is aligned, at equilibrium, with the desired orientation of the reference frame with respect to the body, and which is defined, during the motion, by the following conditions,

$$\int \underset{\sim}{u}\, dm = 0 \quad , \quad \int \underset{\sim}{x} \times \underset{\sim}{u}\, dm = 0 \quad , \tag{20}$$

where $\underset{\sim}{x}$ is the position vector of dm in undeformed configuration,
$\underset{\sim}{u}$ is the deformation vector,

the above vectors being determined when the rotors do not spin with respect to the body (Fig. 2). The deformations can be expressed in terms of a set of n independent variables which are equal to zero at equilibrium. It should be noted that the deformations described by arbitrarily chosen variables will not necessarily satisfy the relations (20). It is then sufficient to add six additional variables and consider (20) as a set of six constraints between the $(n+6)$ variables. As it is assumed that the variables are equal to zero at equilibrium, constant generalized forces will not appear in the equations and the Lagrange multipliers introduced will be of the order of the other variables. Consequently, the linearized system will be equivalent if the constraint equations are linearized. These constraints can then be integrated, six variables can be eliminated and the deformation can be expressed in terms of the n remaining variables β_ν.

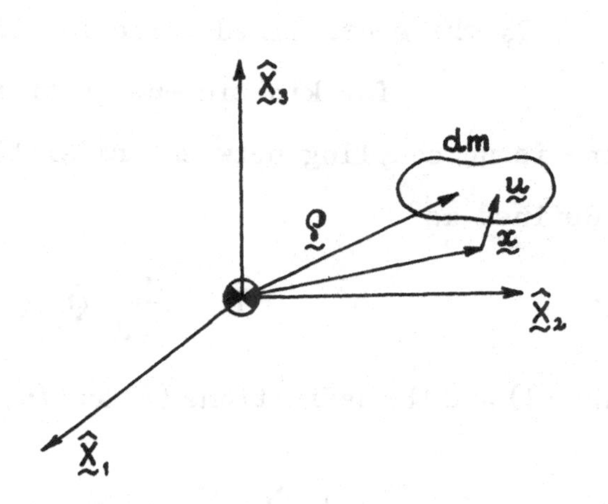

Fig. 2

The equations of deformation are derived from the Lagrange equation in the non cyclic variables β_ν :

$$\frac{d}{dt}\frac{\partial R}{\partial \dot{\beta}_\nu} - \frac{\partial R}{\partial \beta_\nu} + \frac{\partial W}{\partial \beta_\nu} = Q_\nu \qquad (\nu = 1 \dots n), \tag{21}$$

where R is the Routhian function,

W the Rayleigh dissipation function,

Q_ν the generalized force for the variable β_ν .

The kinetic energy of rotation of the system (when there is no coupling between translation and rotation) is a scalar defined as

(22) $$T = \frac{1}{2}\int \dot{\rho} \cdot \dot{\rho} \, dm$$

Using (4) and the definitions (8) and (9), this energy can be written,

$$T = \frac{1}{2}\int \overset{\circ}{\underset{\sim}{\rho}} \cdot \overset{\circ}{\rho} \, dm + \underset{\sim}{\omega} \cdot h + \frac{1}{2} \underset{\sim}{\omega} \cdot \underset{\sim}{J} \cdot \underset{\sim}{\omega} \ ,$$

(23) or

$$T = \frac{1}{2}\int \dot{\rho}^T \rho \, dm + \omega^T h + \frac{1}{2} \omega^T J \omega$$

The internal angular momentum of the system is equal to the sum of $\underset{\sim}{h}_r$ and the internal angular momentum of the corresponding system with "frozen" rotors

(24) $$\underset{\sim}{h} = \int (\underset{\sim}{x} + \underset{\sim}{u}) \times \overset{\circ}{\underset{\sim}{u}} \, dm + \underset{\sim}{h}_r \ .$$

Its relative time derivative is then,

(25) $$\overset{\circ}{\underset{\sim}{h}} = \int (\underset{\sim}{x} + \underset{\sim}{u}) \times \overset{\circ\circ}{\underset{\sim}{u}} \, dm + \overset{\circ}{\underset{\sim}{h}}_r$$

As we consider rigid constant speed rotors, $\overset{\circ}{\underset{\sim}{h}}_r$ is due to the relative angular velocity of the rotor spin axis, $\underset{\sim}{\omega}_d$, with respect to the reference axes or

$$\overset{\circ}{\underset{\sim}{h}}_r = \underset{\sim}{\omega}_d \times \underset{\sim}{h}_r \ . \tag{26}$$

The potential energy in the inverse square field is,

$$V = -\frac{k}{r}m - \omega_0^2 \int \underset{\sim}{\rho} \cdot \underset{\sim}{\rho}\, dm + \frac{3}{2}\omega_0^2 \hat{\underset{\sim}{a}}_1 \cdot \underset{\sim}{J} \cdot \hat{\underset{\sim}{a}}_1 \ , \tag{27}$$

or in matrix form,

$$V = -\frac{k}{r}m - \frac{1}{2}\omega_0^2 \, tr J + \frac{3}{2}\omega_0^2 \theta_1^T J \theta_1 \ , \tag{28}$$

where k is the gravitational constant,

m is the mass of the system,

tr meaning "trace of".

The matrix β will be defined as the $n \times 1$ matrix $\beta = [\beta_1 \dots \beta_n]^T$ and a corresponding matrix B will be defined as

$$B = \begin{bmatrix} \beta & 0 & 0 \\ 0 & \beta & 0 \\ 0 & 0 & \beta \end{bmatrix} \tag{29}$$

As only linear systems are considered, we have to keep quadratic terms in the variables in the expressions of the kinetic and potential energies. The inertia matrix will then have to be developped up to quadratic terms and will be written,

$$J = I + 2B^T\Lambda + B^T \Pi B \tag{30}$$

where I is the inertia matrix of the system at equilibrium given by (19),

Λ and Π are constant matrices functions of the shape of the deformations.

The matrices ω, ω_d will be normalized to ω_o and the operator $(\dot{\ })$ will mean, from now, derivative with respect to the mean anomaly of the orbit $\tau = \omega_o t$.

The matrices h_r and ω_d can also be expressed in terms of the variables β_ν. The variables defining the angular position of the rotors being cyclic variables, the relative angular momentum $\underset{\sim}{h}_r$ is constant in a frame, $\{\hat{\underset{\sim}{Y}}_\alpha\}$, moving with respect to the body reference frame with the relative angular velocity $\underset{\approx}{\omega}_d$.

The frame $\{\hat{\underset{\sim}{Y}}_\alpha\}$ coincides with $\{\hat{\underset{\sim}{X}}_\alpha\}$ at equilibrium and the transformation matrix between $\{\hat{\underset{\sim}{X}}_\alpha\}$ and $\{\hat{\underset{\sim}{Y}}_\alpha\}$ is a function of

(31) $$\{\hat{\underset{\sim}{Y}}_\alpha\} = A(\beta)\{\hat{\underset{\sim}{X}}_\alpha\} .$$

For small deformation the matrix $A(\beta)$ can be written up to quadratic terms in β as

(32) $$A(\beta) = E - \tilde{\psi} - B^T \Psi^T B ,$$

where E is the 3×3 unit matrix, ψ is a 3×1 matrix $\psi = [\psi_1 \ \ \psi_2 \ \ \psi_3]^T$, the 3×3 matrix $\tilde{\psi} = [\varepsilon_{\alpha\gamma\delta}\psi_\gamma]$, $\varepsilon_{\alpha\gamma\delta}$ being the Levi-Civita density and the $3n \times 3n$ matrix

Ψ is a constant.

The components ψ_α are the sum of the successive elementary-rotations along the α-axes and the matrix ψ can be written as

$$\psi = [\xi_1\beta \quad \xi_2\beta \quad \xi_3\beta]^T . \tag{33}$$

Defining as h_0 , the matrix of the components of the internal angular momentum at equilibrium, $h_0 = [J_1 \quad J_2 \quad J_3]^T$, the matrix h_r will be during the motion, equal to

$$h_r = [E + \tilde{\psi} - B^T \Psi B] h_0 . \tag{34}$$

The matrix ω_d can then be written under the form:

$$\omega_d = \dot{\psi} + B^T \Xi \dot{\beta} , \tag{35}$$

or

$$\omega_d = [\xi_1^T\dot{\beta} + \beta^T\Xi_1\dot{\beta} \quad \xi_2^T\dot{\beta} + \beta^T\Xi_2\dot{\beta} \quad \xi_3^T\dot{\beta} + \beta^T\Xi_3\dot{\beta}]^T,$$

where the $n \times n$ matrices Ξ_α and then the $3n \times n$ matrix Ξ, are related to the matrices ψ and Ψ.

From the constraint equations it is seen that the integral $\int(\underset{\sim}{x}+\underset{\sim}{u})\times\mathring{\underset{\sim}{u}}dm$ will not include linear terms in β or $\dot{\beta}$ and up to quadratic terms in β and $\dot{\beta}$ this integral will be written:

$$\int(\underset{\sim}{x} + \underset{\sim}{u})\times\mathring{\underset{\sim}{u}}\,dm = [\hat{\underset{\sim}{X}}_\alpha]^T B^T C \dot{\beta} . \tag{36}$$

It is convenient to write the matrix C under the form;

$$C = \begin{bmatrix} \varepsilon_{\alpha 1 \beta} & \Gamma_{\alpha\beta} \\ \varepsilon_{\alpha 2 \beta} & \Gamma_{\alpha\beta} \\ \varepsilon_{\alpha 3 \beta} & \Gamma_{\alpha\beta} \end{bmatrix} \tag{37}$$

The integral $\frac{1}{2}\int \mathring{\underset{\sim}{\rho}} \cdot \mathring{\underset{\sim}{\rho}}\, dm$ in the expression of T is equal to the sum:

$$\frac{1}{2}\int \mathring{\underset{\sim}{\rho}} \cdot \mathring{\underset{\sim}{\rho}}\, dm = \frac{1}{2}\int \mathring{\underset{\sim}{u}} \cdot \mathring{\underset{\sim}{u}}\, dm + \underset{\sim}{\omega}_d \cdot \underset{\sim}{h}_r + T_r \ , \tag{38}$$

where T_r is the rotor kinetic energy of rotation and is a constant. The integral $\frac{1}{2}\int \mathring{\underset{\sim}{u}} \cdot \mathring{\underset{\sim}{u}}\, dm$ is written up to quadratic terms in $\dot{\beta}$ as:

$$\frac{1}{2}\int \mathring{\underset{\sim}{u}} \cdot \mathring{\underset{\sim}{u}}\, dm = \frac{1}{2}\dot{\beta}^T M_d \dot{\beta} \ , \tag{39}$$

where M_d is referred to as the generalized mass matrix.

It will be assumed that the dissipation function is a quadratic in $\dot{\beta}$,

$$W = \frac{1}{2}\dot{\beta}^T Z \dot{\beta} \ , \tag{40}$$

and that the potential energy of deformation has the form,

$$U_d = \frac{1}{2}\beta^T K \beta + F\beta \ . \tag{41}$$

At equilibrium, the dynamical potential has to be minimum and the equilibrium does not necessarily correspond to the equilibrium of the system outside the gravitational field with zero angular momentum.

The attitude of the frame $\{\hat{\underset{\sim}{X}}_\alpha\}$ with respect to the reference frame $\{\hat{\underset{\sim}{a}}_\alpha\}$ will be described by the rotation angles θ_1 θ_2 θ_3, about 1- 2- and 3-axes respectively. The angular velocity of $\{\hat{\underset{\sim}{X}}_\alpha\}$ can then be expressed in terms of the angular velocity of the reference frame and of the angles θ_α and their time derivatives.

If one defines the matrix x as the vector array

$$x = [\,\theta_1 \;\; \theta_2 \;\; \theta_3 \;\; \beta\,]^T , \tag{42}$$

the linearized equations (13) and (21) can be written in matrix form as

$$M\ddot{x} + G\dot{x} + Kx = -D\dot{x} , \tag{43}$$

where M, K and D are asymmetric matrices,

G is a skew-symmetric matrix.

Further additional equilibrium conditions are obtained.

The matrices M, G, K and D are given by

$$M = \begin{bmatrix} I & 0 \\ 0 & M_d \end{bmatrix} , \qquad D = \begin{bmatrix} 0 & 0 \\ 0 & Z \end{bmatrix} ,$$

$$G = \begin{bmatrix} 0 & -(I_{11}+I_{22}-I_{33}-J_3) & J_2 & 2\Lambda^T_{13}+J_3\xi^T_2-J_2\xi^T_3 \\ I_{11}+I_{22}-I_{33}-J_3 & 0 & \frac{J_1}{2} & 2\Lambda^T_{23}+J_1\xi^T_3-J_3\xi^T_1 \\ J_2 & -\frac{J_1}{2} & 0 & 2\Lambda^T_{33}+J_2\xi^T_1-J_1\xi^T_2 \\ -(2\Lambda_{13}+J_3\xi_2-J_2\xi_3) & -(2\Lambda_{23}+J_1\xi_3-J_3\xi_1) & -(2\Lambda_{33}+J_2\xi_1-J_1\xi_2) & \Gamma_d \end{bmatrix}$$

(44)

$$K = \begin{bmatrix} I_{33}-I_{22}+J_3 & 0 & -\frac{b}{4}J_1 & -(2\Lambda^T_{23}+J_1\xi^T_3-J_3\xi^T_1) \\ 0 & -a(I_{11}+I_{33})+J_3 & bJ_2 & 2a\Lambda^T_{13}+J_2\xi^T_2-J_2\xi^T_3 \\ -\frac{b}{4}J_1 & bJ_2 & b(I_{22}-I_{11}) & -2b\Lambda^T_{21} \\ -(2\Lambda_{23}+J_1\xi_3-J_3\xi_1) & 2a\Lambda_{13}+J_3\xi_2-J_2\xi_3 & 2b\Lambda_{12} & \Pi_d \end{bmatrix}$$

where

$$\Gamma_d = \Gamma^T_{21} - \Gamma_{21} - \Gamma^T_{12} + \Gamma_{12} + 2J_1(\xi_2\xi^T_3 - \xi_3\xi^T_2) +$$

$$+ 2J_2(\xi_3\xi^T_1 - \xi_1\xi^T_3) + 2J_3(\xi_1\xi^T_2 - \xi_2\xi^T_1) + \tag{45}$$

$$+ J_1(\Xi^T_1 - \Xi) + J_2(\Xi^T_2 - \Xi_2) + J_3(\Xi^T_3 - \Xi_3) \quad ,$$

and for a gravity stabilized satellite

$$a = 4 \quad , \quad b = 3 \ ,$$

$$\Pi_d = K + \Pi_{11} + \Pi_{11}^T - \Pi_{33} - \Pi_{33}^T - \frac{1}{2}(\Pi_{22} + \Pi_{22}^T) + \qquad (46)$$
$$+ J_1(\Psi_{13} + \Psi_{13}^T) + J_2(\Psi_{23} + \Psi_{23}^T) + J_3(\Psi_{33} + \Psi_{33}^T) \ .$$

For a spinning gyrostat, the relations (46) become:

$$a = 1 \quad , \quad b = 0 \ ,$$

$$\Pi_d = K - \frac{1}{2}(\Pi_{33} + \Pi_{33}^T) + J_1(\Psi_{13} + \Psi_{33}^T) + \qquad (47)$$
$$+ J_2(\Psi_{23} + \Psi_{23}^T) + J_3(\Psi_{33} + \Psi_{33}) \ .$$

The additional equilibrium conditions are for a gravity stabilized system:

$$J_2\xi_1 - J_1\xi_2 - 2\Lambda_{11} + \Lambda_{22} + 2\Lambda_{33} - F = 0 \ , \qquad (48)$$

for a spinning system:

$$J_2\xi_1 - J_1\xi_2 + \Lambda_{33} - F = 0 \ . \qquad (49)$$

These conditions, together with the rigid gyrostat equilibrium conditions, can be considered as a set of algebraic relations between the system parameters. They can be used to determine the equilibria. A system of two rigid bodies connected by a line hinge was investigated. Once the equilibrium is determined the various matrices can be derived and the stability can be obtained. Systems of several interconnected gyrostats could also be analized by this method.

Stability analysis.

The stability can be determined by the Liapunov method. The Hamiltonian of the system described by the system (43) is equal to

(50) $$H = \dot{x}^T M \dot{x} + x^T K x \ .$$

Using (43), the time derivative along a trajectory of this quadratic function is

(51) $$\dot{H} = - \dot{x}^T D \dot{x} \ .$$

The Hamiltonian can be used as a Liapunov function. The matrix D being semi-definite positive, the positive definiteness of the Hamiltonian is a necessary and sufficient condition for stability when the damping is complete and sufficient condition only, when damping is not complete. Damping is said complete when there is no trajectory (in state space) different from a trivial solution $x \equiv 0$, along which the energy dissipation is identically equal to zero.

The matrix M is positive definite by definition, the Hamiltonian will be positive definite when the matrix K is positive definite. From Sylvester's criterion, the determinants of all the principal minors of K have to be positive to have sta bility. For gravity stabilized systems, the damping is complete when all the variables are coupled to the variables β.

For spin stabilized systems, the damping is not complete. In fact, if initial conditions modify the total angular momentum, the satellite will be in equilibrium for some constant values of θ_1, θ_2 and θ_3. The positive definiteness of K provides only sufficient conditions.

If necessary conditions have to be obtained, one must constrain the system to keep its total angular momentum. These constraints must hold for initial conditions and then also for every point of the trajectory. Their introduction permits the elimination of some variables and new (necessary and sufficient) stability criteria are then obtained.

Particular cases of gravity stabilized gyrostats.

When $J_1 = J_2 = J_3 = 0$, the deformable body does not include internal mass circulation. It is seen from (19) that at equilibrium the principal axes of the body are coinciding with the orbital reference frame.

In this case, when the damping is complete, it is seen that the relation,

$$I_{33} > I_{22} > I_{11} \quad , \tag{52}$$

must be satisfied in order to have (asymptotic) stability. This condition is necessary and the requirement of positiveness for other minors of K may only decrease the stability of the system.

When $J_1 = J_2 = 0$ and $J_3 \neq 0$ the matrix I is also diagonal and the principal axes of the body are coinciding at equilibrium with the orbital reference frame, the vector $\underset{\sim}{h}$ being perpendicular to the plane of the orbit.

Necessary conditions for completely damped systems are

$$(53) \qquad I_{33} - I_{22} + J_3 > 0 \quad , \qquad 4(I_{33} - I_{11}) + J_3 > 0 \; ,$$
$$I_{22} > I_{11}$$

This allows the cases:

$$(54) \qquad \begin{aligned} &a) \quad I_{33} > I_{22} > I_{11} \; , \\ &b) \quad I_{22} > I_{33} > I_{11} \; , \\ &c) \quad I_{22} > I_{11} > I_{33} \; , \end{aligned}$$

depending on the sign and the amplitude of the internal momentum.

When J_3 is positive, the presence of rotors increses the stability region. In fact, in this case the system will always be stable for $I_{33} > I_{22} > I_{11}$.

When the body degenerates to a single rotor with moments of inertia $I_{11} = I_{22}$ and I_{33} , J_3 can be written:

$$(55) \qquad J_3 = I_{33} R \; ,$$

where R is the normalized angular velocity of the satellite, and the first two conditions (53) are then

$$(1 - K)R - K > 0 \ , \tag{56}$$

and

$$(1 - K)R - 4K > 0 \ ,$$

where

(57)

$$K = \frac{I_{22} - I_{33}}{I_{11}}$$

The condition $I_{22} > I_{11}$ is never satisfied. At the limit then, neutral stability is obtained.

When $J_1 = 0$ and $J_2 \neq 0$ the axis $\hat{\underset{\sim}{a}}_1$ is coinciding at equilibrium with one principal axis of the body, say the axis $\hat{\underset{\sim}{a}}_1$.

The inertia matrix (19) can then be written as

$$I = \begin{bmatrix} I_{11} & 0 & 0 \\ 0 & I_{22} & -J_2 \\ 0 & -J_2 & I_{33} \end{bmatrix} . \tag{58}$$

The principal axes of the body, $\hat{\underset{\sim}{y}}_2$ and $\hat{\underset{\sim}{y}}_3$, with respective moment of inertia I_2 and I_3, are located in the $\hat{\underset{\sim}{x}}_2 - \hat{\underset{\sim}{x}}_3$ plane. If the angle between $\hat{\underset{\sim}{x}}_2$ and $\hat{\underset{\sim}{y}}_2$ is θ, the matrix I can be written

$$I = \begin{bmatrix} I_1 & 0 & 0 \\ 0 & I_2\cos^2\theta + I_3\sin^2\theta & (I_2 - I_3)\sin\theta\cos\theta \\ 0 & (I_2 - I_3)\sin\theta\cos\theta & I_3\cos^2\theta + I_2\sin^2\theta \end{bmatrix} . \tag{59}$$

If Φ is the angle between $\hat{\underset{\sim}{x}}_2$ and $\underset{\sim}{h}$ at equilibrium, the vector $\underset{\sim}{h}$ being then in the $\hat{\underset{\sim}{x}}_2 - \hat{\underset{\sim}{x}}_3$ plane, and if h is the norm of $\underset{\sim}{h}$, one has the well-known relation between Φ and θ

$$2h\cos(\theta + \Phi) = (I_3 - I_2)\sin^2\theta \tag{60}$$

The positiveness of K implies:

$$(I_3 - I_2)(\cos^2\theta - \sin^2\theta) + [h^2 - (I_2 - I_3)^2\sin^2\theta\cos^2\theta]^{1/2} > 0 \,,$$

$$4(I_3 - I_2)\cos^2\theta + 4(I_2 - I_1) + [h^2 - (I_2 \;\; I_3)^2\sin^2\theta\cos^2\theta]^{1/2} > 0 \,,$$

$$(I_2 - I_3)\cos^2\theta + (I_3 - I_1) > 0 \,, \tag{61}$$

$$[(I_2 - I_3)\cos^2\theta + (I_3 - I_1)]\{4(I_3 - I_2)\cos^2\theta + 4(I_2 - I_1) +$$

$$+ [h^2 - (I_2 - I_3)^2\sin^2\theta\cos^2\theta]\}^{1/2} - 3(I_2 - I_3)^2\sin^2\theta\cos^2\theta > 0 \,,$$

which are the conditions of rigid gyrostats. The positiveness of other principal minors of K may only decrease the stability region.

It must be noted that the third condition implies;

$$\cos^2\theta > \frac{I_1 - I_3}{I_2 \;\; I_3} \,. \tag{62}$$

The equality being obtained when

$$J_2 = \sqrt{(I_1 - I_3)(I_2 - I_1)} \tag{63}$$

It is seen that (53) is always satisfied when I_3 is an intermediate axis of inertia.

When $J_2 = 0$ and $J_1 \neq 0$, the axis $\hat{\underset{\sim}{a}}_2$ is a principal axis at equilibrium.

The inertia matrix may now be written:

$$I = \begin{bmatrix} I_{11} & 0 & -\frac{J_1}{4} \\ 0 & I_{22} & 0 \\ -\frac{J_1}{4} & 0 & I_{33} \end{bmatrix} . \tag{64}$$

The other two principal axes $\hat{\underset{\sim}{y}}_1$ and $\hat{\underset{\sim}{y}}_3$ with moment of inertia I_1 and I_3 respectively are located in the $\hat{\underset{\sim}{x}}_1 - \hat{\underset{\sim}{x}}_3$ plane. Defining the angle between $\hat{\underset{\sim}{x}}_1$ and $\hat{\underset{\sim}{y}}_1$ as θ, the inertia matrix is now equal to

$$I = \begin{bmatrix} I_1\cos^2\theta + I_3\sin^2\theta & 0 & (I_3 - I_1)\sin\theta\cos\theta \\ 0 & I_2 & 0 \\ (I_3 - I_1)\sin\theta\cos\theta & 0 & I_1\sin^2\theta + I_3\cos^2\theta \end{bmatrix} . \tag{65}$$

At equilibrium, the vector $\underset{\sim}{h}$ is located in the $\hat{\underset{\sim}{x}}_1 - \hat{\underset{\sim}{x}}_2$ plane and the angle Φ between $\underset{\sim}{h}$ and $\hat{\underset{\sim}{y}}_1$ is related to the angle θ by the relation

$$\frac{h}{2}\cos(\theta + \Phi) = (I_1 - I_3)\sin 2\theta \ . \tag{66}$$

The necessary conditions for stability for the corresponding rigid body are then

$$(I_3 - I_1)\cos^2\theta + (I_1 - I_2) + [h^2 - 16(I_3 - I_1)^2\sin^2\theta\cos^2\theta]^{1/2} > 0,$$

$$4(I_3 - I_1)(\cos^2\theta - \sin^2\theta) + [h^2 - 16(I_3 - I_1)^2\sin^2\theta\cos^2\theta]^{1/2} > 0,$$

$$(I_2 - I_3) + (I_3 - I_1)\cos^2\theta > 0, \tag{67}$$

$$[(I_3 - I_1)\cos^2\theta + (I_2 - I_1)]\{(I_3 - I_1)\cos^2\theta + (I_1 - I_2) +$$

$$+ [h^2 - 16(I_3 - I_1)^2\sin^2\theta\cos^2\theta]^{1/2}\} - 3(I_2 - I_3)^2\sin^2\theta\cos^2\theta > 0,$$

and once more the conditions on the other principal minors of K may only decrease the stability region.

The third condition is now,

$$\cos^2\theta > \frac{I_2 - I_3}{I_1 - I_3}, \tag{68}$$

the equality sign being obtained when

$$J_1 = 4\sqrt{(I_1 - I_2)(I_2 - I_3)} \ . \tag{69}$$

This is also a particular value for which there is only partial stability.This is to be expected the point into consideration being part of a continuum of equilibria.

All the stability criteria for rigid gyrostats associated with internal damping were obtained as particular cases. Further, the system presented here includes a physical damping and then is not a pure idealization. It was shown that, in all the cases, the presence of deformations may only decrease the attitude stability of gyrostats as well as satellites without internal circulation of mass.

Contents

Printed in the United States
By Bookmasters